FRACTIONATION

Edited by **Hassan Al- Haj Ibrahim**

Fractionation
http://dx.doi.org/10.5772/intechopen.72270
Edited by Hassan Al- Haj Ibrahim

Contributors
Hassan Al- Haj Ibrahim, Juan Pablo Salazar Giraldo, Yew Keong Choong, Kavithambigai Ellan, Xiang-Dong Chen, Shaiful Azuar Mohamad, Nelio Teixeira Machado, Luiz Eduardo Pizarro Borges, Marcelo Costa Santos, Sergio Duvoisin, Caio Campos Ferreira, Douglas Alberto Rocha De Castro, Anderson Pereira, Wenderson Santos, Haroldo Jorge Da Silva Ribeiro, Lauro Henrique Hamoy Guerreiro, Márcio Cordeiro, Jose Otavio Carrera Silva Junior

First published in London, United Kingdom, 2019 by IntechOpen
IntechOpen is the global imprint of INTECHOPEN LIMITED, registered in England and Wales, registration number: 11086078, The Shard, 25th floor, 32 London Bridge Street
London, SE19SG – United Kingdom
Printed in Croatia

British Library Cataloguing-in-Publication Data
A catalogue record for this book is available from the British Library

Additional hard copies can be obtained from orders@intechopen.com

Fractionation, Edited by Hassan Al- Haj Ibrahim
p. cm.
Print ISBN 978-1-78984-964-6
Online ISBN 978-1-78984-965-3

For EU product safety concerns:
IN TECH d.o.o., Prolaz Marije Krucifikse Kozulić 3, 51000 Rijeka, Croatia,
info@intechopen.com or visit our website at intechopen.com.

Meet the editor

After graduating in Fuel Engineering from Leeds University, England, I was appointed British Coke Research Association Fellow, and in 1973 I was awarded my Ph.D. After graduation I worked at an oil refinery and in other research establishments in the UK and Syria. In 1980 I spent an academic year at the Technical University of Aachen, and in 1990 I was awarded the Fulbright scholarship and spent an academic year at the University of Pittsburgh. For the last 40 years I have been a professor of petroleum refining at Al-Baath University, Syria, and Director of Quality Assurance (since 2003).
My publications include research papers on chemical engineering and on quality assurance in education, literature, and general culture. I have also written several textbooks on petroleum refining, metallurgy, sources of energy, and a dictionary of petroleum terms.

Contents

Chapter 1

Introductory Chapter: Fractionation

Hassan Al-Haj Ibrahim

Additional information is available at the end of the chapter

http://dx.doi.org/10.5772/intechopen.78050

1. Introduction

Separation processes whereby different properties are used in the separation of mixtures of different compositions are some of the most important chemical engineering processes. The mixtures to be separated may be mixtures of isotopes, molecules, polymers or cells. The different properties on which the separation process is based may be chemical or physical including such properties as chemical reactivity, solubility, molecular size, electrical charge and change-of-phase temperatures such as boiling and freezing points.

The application of separation processes ranges from such operations as water purification to refining of products and separation of raw materials. Separation processes account for 40–90% of the capital investment in industrial establishments. It has been variously estimated that the capital investment in separation equipment is 40–50% of the total for a conventional fluid processing unit [1].

There are many different separation techniques which may be broadly classified into processes of mechanical separation and separation by diffusion. Mechanical separation techniques may be based on particle size, density and electrical or magnetic mobility. Separation by diffusion includes chromatographic separation, extraction and fractionation. Of the many and varied separation processes available, fractionation plays a major part, and capital investment in fractionation equipment and processes may form a significant fraction of industrial processing investment.

Fractionation, in the most general sense, could mean any process whereby a mixture is separated into different components or fractions. Fractionation in this general sense, however, would include diverse methods and techniques that may have little in common. Such methods and techniques may be broadly grouped in four classes, namely fractionation proper, general separation, analytical separation and purification. It is important to note, however, that this is not an exclusive classification as some methods may have characteristics belonging to more than one of these classes. Chromatography, for example, may be either analytical or separatory:

1. Fractionation proper in the restricted commonly accepted sense, sometimes referred to as thermal separation, may be defined as any physical fractional or separation process involving a phase transition in which a mixture is separated in a single run into two or more fractions with variable compositions. The mixture to be separated could be a homogeneous mixture such as a solution or a heterogeneous mixture such as a mixture of solid and liquid. The mixture to be separated may also be gaseous, liquid or solid. The fractions are separated based on differences in a specific property of the individual components such as boiling, freezing or melting points. The phase changes involved could be condensation for vapours and gases (dephlegmation), boiling and evaporation (distillation) or freezing and crystallisation (freeze distillation and solution crystallisation) for liquids, and melting or sublimation for solids. Most of these processes find application on industrial scales but some processes such as fractionation by sublimation are restricted to laboratory-scale procedures.

The general characteristics of fractionation processes may thus be summarised as follows:

1. Fractionation processes are processes of separation of mixtures of liquids, gases or solids into their components.
2. Fractional separation is based on differences in a specific property of the individual components.
3. Fractionation processes are physical processes, where phase transition is involved.
4. They are carried out in single runs.

Examples of fractionation processes include (**Table 1**):

1. Dephlegmation, where mixtures of vapours are separated by difference in their condensation point.
2. Fractional distillation, where mixtures of liquids and gases are separated by difference in boiling point.
3. Fractional freezing, where mixtures of liquids are separated by difference in freezing point.
4. Fractional melting, where mixtures of solids are separated by difference in melting point.
5. Isotope fractionation, where mixtures of isotopes are separated by difference in density during phase transition.

2. The term 'fractionation', however, may sometimes be used in a more general sense to refer to any separation process that may involve no phase change and may not therefore be classified as a fractionation process properly speaking. Such separation processes include:

2.1. Solvent or clean fractionation, where a mixture of an organic solvent and water is used to achieve clean separation.

http://dx.doi.org/10.5772/intechopen.78050

Fractionation process		Mixture	The specific property	Area of application
Dephlegmation		Vapours	Condensation point	Oil refining
Fractional distillation		Liquids and vapours	Boiling point	Petroleum and petrochemical industries
Fractional crystallisation = fractional freezing	Partial crystallisation	Liquids	Freezing point	Fractionation of food Upgrading of alternative fuels Obtaining very pure substances.
	Solution crystallisation	Liquids	Solubility	
Fractional melting		Solids	Melting point	
Isotope fractionation			Density	

Table 1. Fractionation processes.

2.2. Thermal diffusion, where mixtures of gases or liquids are separated by difference in density.

2.3. Centrifugation, which is a process where substances of different densities in a heterogeneous mixture are separated by sedimentation using a centrifuge.

2.4. Gaseous diffusion, where mixtures of gases are separated by difference in molecular weight.

2.5. Chromatography where fractionation of components from a solution mixture takes place by difference in affinity between the stationary and the mobile phases.

3. The term 'fractionation' may also be used sometimes to refer to an analytical technique and not to a separation process. Examples of such use include bioassay-guided fractionation, analytical chromatography and geochemical fractionation.

Bioassay is an analytical method to determine the concentration or potency of a substance by its effect on living cells or tissues. In bioassay-guided fractionation, extracted components are separated based on differences in their physicochemical properties and biological activity. In pharmacognosy, a pure chemical agent from natural origin is isolated by bioassay-guided fractionation.

Geochemical fractionation is a laboratory procedure in which chemical extraction is used to determine fractions of metals in soils and sediments and their susceptibility to release into an aqueous environment.

4. Finally, fractionation may refer to a purification process, where a desired component is extracted from a mixture and purified as, for example, in the pharmaceutical industry, where natural and synthetic drugs are separated and purified to meet health needs. Surface Active Foam Fractionation, a new technology developed by scientists from a privately-owned Australian company, may be used to remove certain contaminants from water. In this

technology, air is introduced into the column of contaminated water through a diffuser, and the resulting bubble column rises to produce contaminant-rich foam on the water surface, which is extracted using a uniquely-designed vacuum extraction system.

Such purification processes, however, may not be considered as separation or fractionation processes per se.

Fractionation in the restricted commonly accepted sense as a separation process properly speaking is widely used in the chemical industries including food and pharmaceutical industries and in petroleum and petrochemical industries in particular. It is normally used to separate a non-usable crude mixture into separate usable fractions as, for example, in oil refineries where crude oil is fractionated into useful petroleum products such as fuels, lubricants and chemical raw materials. In biorefineries, lignocellulosic raw materials are fractionated into their polymeric and non-polymeric constituents and new routes developed for their chemical and biochemical conversion into high-value-added products.

1.1. Dephlegmation

Dephlegmation is a separation process where mixtures of vapours are separated by difference in their condensation point. In chemistry, the noble gases are separated from one another by fractionation, making use of their different condensation temperatures.

1.2. Fractional distillation

Fractional distillation is a distillation process where mixtures of liquids and gases are separated by difference in boiling point. Fractional distillation processes include topping, rectification, exhausting, stripping, vacuum distillation, stabilization and extractive distillation.

Distillation as a separation process is used in general to separate a fluid mixture of two (binary) or more (multi-component) substances into its component parts. In most cases, the components to be separated are miscible liquids with different volatilities and boiling points. This separation process is a thermal unit operation that utilizes the differences of vapour pressure to produce the separation. In this process, the vapour or liquid mixture is heated whereby the more volatile components are evaporated, condensed and allowed to drip or drip apart, that is, distil or destillare, as it was originally called in Latin. It is from this fact of 'dripping' that the name of the distillation process was derived [1].

In general, distillation is favoured over other separation techniques when large rates are desired of products that are thermally stable with a relative volatility greater than 1.2 and where no extreme corrosion, precipitation, sedimentation or explosion issues are present [2].

The pre-eminence of distillation for the separation of fluid mixtures is fundamental for both kinetic and thermodynamic reasons [3]. It may also be the most economical process for the separation and production of high-purity products. In 1992, Darton estimated the worldwide throughput of distillation columns at 5 billion tonnes of crude oil and 130 million tonses of *petrochemicals* per year [1].

Fractional distillation is a special type of distillation, and as a separation technique, is much more effective than simple distillation and more efficient. Fractional distillation is equivalent in effect to a series of distillations, where separation is achieved by successive distillations or repeated vaporization-condensation cycles. Each vaporization-condensation cycle makes for an equilibrium stage, commonly known as a theoretical stage. A number of such theoretical stages may be required for the efficient fractionation and separation of the vapour or the liquid mixture. The McCabe-Thiele method is a graphical method that may often be used to calculate the required number of theoretical stages. These theoretical stages will then have to be converted to actual plates or an equivalent packed height depending upon the separation efficiency for a particular service. As a separation technique, fractional distillation is much more effective than simple distillation and more service [1].

In fractional distillation, the components are separated through continuous heat and mass transfer between counter-current streams of a rising vapour and a descending liquid. As in all thermal separation processes, the motive force for the separation is the drive towards thermodynamic equilibrium between the different phases (VLE or vapor liquid equilibrium). This equilibrium is continuously disturbed by the mixing of the colder descending liquid and the hotter rising vapour, where the more volatile components of the descending liquid are vaporized and the less volatile components of the rising vapour are condensed and the driving force for the separation process is thereby maintained. The concentration of the lighter components will be greater in the vapour phase and conversely the concentration of the heavier components will be greater in the liquid phase, and only in the case of pure components or azeotropic mixtures will the equilibrium composition be the same in both phases [1].

In recent years, fractional distillation has become one of the most important unit operations in chemical engineering industries. It is by far the most common specialized separation technology. In the USA, fractionators are used in 90–95% of the separations [4].

In petroleum refineries, in particular, atmospheric fractionation of crude oil, variously known as topping, is the first and the most important process of the series of oil refinery operations [5]. The crude oil feedstock, which is a very complex multi-component mixture, is separated into a number of products or fractions, with each product or fraction being composed of groups of compounds within a relatively small range of boiling points. These straight-run products or fractions are the origin of the term fractional distillation or fractionation. Fractional distillation, however, is not used in the petroleum-refining industry only, but is also used in other chemical, petrochemical, beverage and pharmaceutical industries and in natural gas-processing plants [1]. In chemistry, fractional distillation may be used for the separation of a mixture of isotopes, where the molecules of the fraction having the lower boiling point (the lighter isotopes) tend to concentrate in the vapour stream and are collected.

Fractional distillation, though widely used, is one of the most energy-intensive operations. In fact, distillation may be the largest consumer of energy in petroleum and petrochemical processing. Of the total energy consumption of an average unit, the separation steps account for about 70% and distillation alone can consume more than 50% of a plant's operating energy cost [6]. Fractional distillation accounts for about 95% of the total distillation consumption [7].

Replacing fractional distillation by other separation techniques in such industries as petroleum and petrochemical industries is, however, highly unlikely despite its energy intensiveness [1].

Extractive distillation is a special type of distillation which is used for the separation of components with similar evaporation points. Such components cannot be separated by simple distillation, because their volatility is nearly the same, causing them to evaporate at nearly the same temperature at a similar rate, making normal distillation impractical. In extractive distillation, a separation solvent is used, which is generally non-volatile, has a high boiling point and is miscible with the mixture, but does not form an azeotropic mixture. The solvent interacts differently with the components of the mixture, thereby causing their relative volatilities to change. This enables the new three-part mixture to be separated by normal distillation. The original component with the greatest volatility separates out as the top product. The bottom product consists of a mixture of the solvent and the other component, which can again be separated easily.

1.3. Fractional freezing

Fractional freezing, also known as fractional crystallisation, is a process where mixtures of liquids are separated by difference in freezing point. It was further observed, when reported data on crystallisation were analysed, that fractionation occurs according to molecular weight. Fractional crystallisation is generally used to produce ultra-pure solids or to recover valuable products from waste solutions and to concentrate liquids. In geology, a process of fractional freezing is known to occur during the metamorphism and structural change of rocks.

Fractional freezing is a simple process that does not require sophisticated equipment or complicated procedures, and it can be used on an industrial scale. Fractional freezing can be used in particular to concentrate heat-sensitive liquids, such as fruit juice concentrates, as it does not involve heating the liquid (as happens during evaporation). However, it is not always possible with this method to produce pure products or to eliminate impurities completely, elimination of impurities being dependent on the nature of these impurities. Furthermore, part of the desired component will always be lost in the discarded material.

Fractional freezing may be carried out by either of two processes, namely

1. Partial crystallisation, also called normal freezing or progressive freezing, where mixtures of liquids are separated by difference in freezing point. This is done by cooling and partial freezing of the liquid and removing the frozen material that is poorer in the desired material than the liquid portion left behind. This is sometimes known as freeze distillation because the resulting enrichment of the liquid portion parallels enrichment by true distillation, where the evaporated and re-condensed portion is richer in the desired lighter material than the liquid portion left behind.

Fractional freezing by partial crystallisation can be used as a simple method to reduce the gel point of biodiesel and other alternative diesel fuels and to increase the alcohol concentration in fermented alcoholic beverages.

The separation of a mixture of water and alcohol is a well-known example of partial crystallisation. Freezing in this case begins at a temperature significantly below 0°C. The first material to freeze is a dilute solution of alcohol in water. The frozen material, while always poorer in alcohol than the liquid, becomes progressively richer in alcohol. The liquid left behind, on the other hand, becomes increasingly richer in alcohol, and, as a consequence, further freezing would take place at progressively lower temperatures. The degree of concentration of the alcohol in the liquid will depend on the final cooling temperature. But even if the final temperature is somewhat below the freezing point of alcohol, there will still be alcohol and water mixed as a liquid, and at some still lower temperature, the remaining alcohol-and-water solution will freeze.

The best-known freeze-distilled beverages are applejack, made from hard cider, and ice beer (Eisbock). In the olden days, barrels of beer were left outside in the winter months and the formed ice, mainly composed of water, would be periodically removed, and this would continue until no more ice is formed. In effect, the presence of ethanol lowers the melting point of water in a solution of water and ethanol, and the solidified phase would contain lesser amounts of ethanol than the liquid phase which will be gradually enriched to give eventually ice beer.

In practice, while not able to produce an alcohol concentration comparable to distillation, partial crystallisation can achieve some concentration with far less effort than any practical distillation apparatus would require. The danger of freeze distillation of alcoholic beverages is that, unlike heat distillation where the methanol and other impurities can be separated from the finished product, freeze distillation does not remove them. Thus, the ratio of impurities may be increased compared to the total volume of the beverage. This concentration may cause side effects to the drinker, leading to intense hangovers and a condition known as 'apple palsy', although this term has also been used simply to refer to intoxication, especially from applejack.

Partial crystallisation may also be used to desalinate sea water which contains principally sodium chloride. Because sodium chloride lowers the melting point of water, the salt in sea water tends to be forced out of pure water while freezing. This is sometimes referrd to as brine rejection.

A fractional freezing technique that permits the freezing of large quantities of solutions is the so-called slow fractional freezing. This is a costly procedure, however, that requires the use of rather expensive equipment. In one such technique, a cylinder containing the liquid is cooled where freezing of the solution starts at the bottom of the cylinder and the frozen solid rises slowly towards the surface of the cylinder. The freezing may be stopped at the desired point for the separation of the solid from the liquid. In another technique that allows better control of the freezing rate, a spherical container is used where freezing starts at the surface of the sphere and moves slowly towards its centre leaving the desired purified liquid at the centre of the sphere surrounded by the frozen solid from which it can be extracted.

2. Solution crystallisation, where chemical substances are fractionated based on difference in solubility at a given temperature. This is done by the addition of a dilute solvent to the liquid mixture and cooling and partial freezing of the liquid and removing of frozen material. If the desired dissolved substance to be separated is the most soluble substance at the specific cooling temperature, the frozen material will be poorer in that desired substance. On the other hand, the liquid portion left behind will be richer in the desired substance. The proportion of the components in the frozen

material and the liquid portion will depend on their solubility products. If the solubility products are very similar, a cascade process will be needed to effect a complete separation.

1.4. Fractional melting

Fractional melting is a fractionation process where mixtures of solids are separated by difference in melting point. This can be done by partial melting of the solid. Fractional melting may be used to desalinate sea water as sea ice, frozen salt water, when partially melted, leaves behind ice that is of a much lower salt content. Fractional melting may also be used to separate waxes into fractions of different melting points which can then be recombined.

1.5. Isotope fractionation

Isotope fractionation occurs during a phase transition as a result of the different distribution of the lighter and heavier isotopes between the liquid and vapour phases. For example, when water vapour condenses, the heavier water isotopes (^{18}O and ^{2}H) become enriched in the liquid phase while the lighter water isotopes (^{16}O and ^{1}H) tend towards the vapour phase.

Other separation processes that are referred to sometimes as fractionation processes although they do not involve in general any phase changes:

1.5.1. Clean or solvent fractionation

In clean fractionation, a mixture of an organic solvent and water is used to achieve clean separation. This fractionation process may be used for upgrading biomass feedstocks for a biorefinery by separating their three main components, viz. cellulose, hemicelluloses and lignin, into pure streams for conversion into value-added products.

1.5.2. Thermal diffusion

In thermal diffusion, natural convection is utilized in the separation of mixtures of liquids or gases by difference in density. In a region with a temperature gradient, the lighter molecules tend to concentrate in the hot parts of the region and the heavier molecules tend to concentrate in the cold parts. Moreover, hotter molecules tend to rise as a result of natural thermal convection and cooler molecules tend to fall, effecting thereby a separation between the lighter and heavier components of the mixture. As no phase change is necessarily involved in thermal diffusion, this process is not normally considered a fractionation process.

A simple form of a thermal diffusion apparatus consists of a tall vertical tube with a wire electrically heated to a specified temperature running down its centre, producing a temperature gradient between the centre and the wall of the tube. The heavier molecules tend to concentrate in the outer portions of the tube, and the lighter molecules to concentrate towards the centre. At the same time, because of thermal convection, the gas or liquid near the wire tends to rise, and the cooler outer gas or liquid tends to fall. The overall effect is that the heavier molecules collect at the bottom of the tube and the lighter at the top.

Thermal diffusion is often used for the separation of mixtures of isotopes.

1.5.3. Centrifugation

Centrifugation is a process where substances of different densities in a heterogeneous mixture are separated by sedimentation using a centrifuge. Centrifugation, however, though a separation process, may not be considered a fractionation process properly speaking as it involves no phase change.

Centrifugation is used in industrial and laboratory settings, particularly in the chemical and food industries and in biological research. Centrifugation is also the most common method used for uranium enrichment, relying on the slight mass difference between atoms of U238 and U235 in uranium hexafluoride gas. In isotope fractionation, the centrifugal force acts more strongly on the heavy molecules than on the light ones, increasing the concentration of the heavy isotopes in the outer region.

In cell fractionation, also called cell lysis, cell components are separated by difference in density while preserving individual functions of each component. With this methodology, the different individual cell components may be examined and their biochemical synthesis and capacity investigated. Cell fractionation is being increasingly applied in research activities in the pharmaceutical industry and disease investigation.

In blood fractionation, blood is separated into its three major components, plasma, buffy coat and erythrocytes. These separated components can be analysed and often further separated. In plasma protein fractionation, the inherent differences of each protein are used and the temperature or the acidity of the plasma is changed so that proteins that are normally dissolved in the plasma fluid become insoluble and can be collected by centrifugation. In cold alcohol fractionation or ethanol fractionation, which is one of the very effective ways for carrying out plasma protein fractionation, alcohol is added to the plasma with simultaneous cooling. Plasma fractionation is used in the prevention and treatment of life-threatening diseases caused by trauma, immunologic disorders and infections. It is also used in treating protein deficiency in health-care applications.

1.5.4. Gaseous diffusion

In gaseous diffusion, the difference in the rate of diffusion of gases of different molecular weight through a porous barrier is utilized to effect separation. As the rate of diffusion of a gas is inversely proportional to the square root of its mass, light atoms diffuse through a porous barrier faster than heavier atoms.

Gaseous diffusion was used in the early forties of the last century for the separation of uranium isotopes. As the difference in weight between uranium-235 and uranium-238 is slight, however, thousands of separation stages would be required to produce 99% uranium-235 from natural uranium.

1.5.5. Chromatography

Chromatography is a technique for the separation of mixtures, where the mixture to be separated is dissolved in a fluid called the mobile phase (the eluent), which carries it through a structure holding another material called the stationary phase. The mobile phase may be a liquid (LC and capillary electrochromatography (CEC)), a gas (GC) or a supercritical fluid (supercritical-fluid chromatograph SFC). The stationary phase may be a thin layer of an adsorbent like silica gel, alumina or cellulose on a flat, inert substrate in thin layer chromatography (TLC) or a strip of chromatography paper in paper chromatography. Furthermore, the stationary phase may be as or on a plane in planar chromatography or the bed of the stationary phase may be in a tube in column chromatography.

The various components of the mixture to be separated travel at different speeds causing them to separate. Fractionation of the mixture components (the analytes or eluites) takes place by difference in affinity between the stationary and the mobile phases. Several mechanisms may be applied for the separation of the analytes such as ion exchange (in ion exchange chromatography) and size exclusion (in size exclusion chromatography, also known as gel permeation chromatography or gel filtration chromatography). In ion exchange chromatography, the analytes are separated based on their respective charges. In size exclusion chromatography, molecules are separated according to their size (or hydrodynamic diameter or volume).

Chromatography may be preparative or analytical, but only preparative chromatography is a form of separation. In chromatography, however, no phase transition is involved and even preparative chromatography may not therefore be considered a true fractionation process.

1.5.6. Other processes of separation

Other processes of separation that are mainly used for isotope separation include electromagnetism and laser separation. In electromagnetism, first used for the production of uranium-238, an ionic beam obtained from a uranium compound was passed through a magnetic field. Because the radius of the curvature of the path of the ions deflected by the beam depends on the mass of the ion, ions of different mass complete their path at different positions, and the uranium isotopes were appreciably separated. In laser separation, the isotope mixture is first vaporized and its atoms are selectively excited and ionized by an accurately tuned laser beam and the desired isotope is thus separated out.

Author details

Hassan Al-Haj Ibrahim

Address all correspondence to: hasahi123@hotmail.com

Al-Baath University, Homs, Syria

References

[1] Al-Haj Ibrahim H. Design of fractionation columns. In: Bennett K, editor. Matlab: Applications for the Practical Engineer. Intech; 2014. pp. 139-171

[2] Kolmetz K, et al. Optimize Distillation Column Design for Improved Reliability in Operation and Maintenance, 2nd Best Practices in Process Plant Management, Kuala Lumpur, Malaysia; March 14-15 2005

[3] Kister HZ. Distillation design. New York: McGraw-Hill Book Co.; 1992

[4] Humphrey JL, Keller GE. Separation Process Technology. New York: McGraw-Hill; 1997

[5] Al-Haj Ibrahim H, Daghistani N. Petroleum Refinery Engineering (in Arabic). Vol. 3. Homs, Syria: Al-Baath University; 2008

[6] Distillation Column Selection and Sizing, KLM Technology Group, 2011. Available from: www.klmtechgroup.com

[7] Chen H, Lin Y. Case studies on optimum reflux ratio of distillation towers in petroleum refining processes. Tamkang Journal of Science and Engineering. 2001;**4**(2):105-110

Chapter 2

Application of the Geochemical Fractionation of Metals in Sediments for Environmental Analysis of a Water Reservoir. Case Riogrande Ii (Antioquia - Colombia)

Juan Pablo Salazar Giraldo

Additional information is available at the end of the chapter

http://dx.doi.org/10.5772/intechopen.76223

Abstract

The geochemical fractionation of metals in soils and sediments corresponds to a technique to evaluate the levels of contamination and their probability of transfer to bodies of water and biota. For environmental studies in water reservoirs, the results of geochemical fractionation added to physicochemical analysis of water, can define the environmental conditions of metal release. This chapter briefly presents the concept and some fractionation techniques, with emphasis on the BCR methodology, in conjunction with other analyzes of water from the bottom of the reservoir to evaluate the dynamics of Mn mobilization in the Riogrande reservoir in Colombia, as example of practical application of Geochemical Fractionation. The highest proportions of Mn in the sediments of the Riogrande II reservoir were found in the exchangeable fraction and associated with carbonates, however the diffraction analysis did not find carbonated phases. It was concluded that the Mn in the water of the bottom of the Riogrande II reservoir originated especially by processes of desorption of Mn, in addition to reductive dissolution of oxyhydroxides.

Keywords: geochemical fractionation, sediments, BCR protocol, metal mobility, Mn dynamics in sediment-water interface

1. Introduction

The geochemical fractionation methodology allows establishing the proportions of groups, fractions or different forms of metals associated with materials such as soils, sediments and sludge, however it is necessary to combine these results with other analyzes such as mineralogical analysis, diffractometry, geoavailability, bioaccessibility and bioavailability analysis, for a complete environmental study. In this chapter of the book "Fractionation", an explanation is given for the concept of Geochemical Fractionation, its importance, the BCR methodology and its application to the release of metals such as Mn in sediments, applied to the Riogrande II Water Reservoir in Colombia.

2. The problem of water contamination in reservoirs by the release of heavy metals from sediments

2.1. Metals in sediments in water reservoirs and lakes

The sediments of the reservoirs, lakes and oceans reflect a recent environmental history, since they are geological records of climatic changes, geodynamic processes, land uses and especially human activity that positively or negatively impacted the environment. The sediments are characterized by the storage of important concentrations of heavy metals and xenobiotic substances due to their high adsorbent capacity, the product of a number of functional groups that allow them to form surface complexes. However, the potential for release and contamination of water from sediments depends on: total concentrations of all substances, environmental conditions of the bottom water (pH, ORP-Eh and organic matter) and forms in which metals are found in the sediments (metals fractions).

Metals have several physicochemical properties and especially different valences, which allows them to be found in different chemical forms or fractions in solid materials such as soil, sediments and mining mud. Some of these forms or geochemical forms are more available than others; so, their release and mobilization to an aqueous phase occurs with slight changes in pH, ORP-Eh [1], temperature and electrical conductivity. The water quality of a reservoir is affected by the presence of dissolved metals and by the conditions and mechanisms of release from the soil or sediments [2], and it has been found that the physicochemical and biological properties and conditions of water quality are affected by external factors such as topography, climate and especially human activity [3], this is how, in general, the most mobile metal form, such as the interchangeable form or fraction, are related to released metals by mining or industrial activity. The sediment is normally located in the lower layer of hypolimnion, where the anoxic and reducing conditions cause several soluble compounds to increase their concentration, especially in the interstitial water through dissolution or reduction and diffuse to the lower layer of water. Temperature as the factor of thermal stratification is an abiotic factor that controls important processes for organisms, likewise it affects the physicochemical properties and other abiotic factors [4, 5]. Within the geochemical fractions or metals forms, when particles settle in a reservoir, a large part of the metal is in an adsorbed form, however in the Hypolimnion, the strongly reducing conditions can dissolve the oxides and consequently release the metals towards the water column [6].

The anaerobic zones, and specifically the deep water masses of the lakes and lagoons, present physical and chemical properties completely different to the more superficial waters, in addition to the temperature. Two of the physicochemical parameters that define ecosystem conditions correspond to dissolved oxygen as to the redox potential, which as it deepens in the water column decreases and forms two profiles similar to the temperature, with their corresponding points of inflection: oxycline and redoxcline, which can occur both in the water column, as in the water-sediment limit, however the abrupt decrease of the potential in the redoxcline is not due to the depletion of oxygen, but due to the appearance of reducing substances such as hydrogen sulfide (H_2S), [7]. In general, in sediments under anoxic conditions, oxidizable metals forms predominate (reduced species), associated with organic and residual matter, while reducible (oxides) and interchangeable metals forms tend to be unstable and are easily solubilized [8]. Metals have a differential behavior in lentic or semilentic bodies of water where epilimnion oxygen-rich conditions predominate. It forms oxidized species (reducible metal forms), which tend to be insoluble and precipitate forming a sediment rich in oxidized solid phases (M^{3+} or M^{4+}), likewise other minerals such as clays with their negative charge, transport on their surface metals, forming complexes of external surface, which can also be submerged in the reservoir. At the bottom of the hypolimnion and sediment, the anoxic and reducing conditions, as well as biological processes, release it in the form of reduced (M^{2+}), complex and soluble species, whose oxidation process is very sensitive to heterogeneous and homogeneous catalysis, and it is dependent on pH and dissolved oxygen. This process is accelerated by the presence of microbial catalysis through different stages; it has been found that in the sediment, the movement of ions such as Mn and Fe towards the water column may be coupled with an eventual and possible H_2S formation [9, 10].

The redox conditions can influence the behavior of the trace metals in sediments and affect the proportions of the metals forms, either directly or indirectly through changes in the oxidation states of the ligands capable of complexing the metal; for example, changes in the redox conditions can cause the reductive dissolution of mineral species that have some adsorbed metals (oxyhydroxides that can be reduced and desorb metal ions): ORP-Eh values higher than +414 mV are considered oxic and oxidizing, the probability of metal release is low; ORP-Eh values between +414 and + 120 mV indicate sub-toxic, moderately reducing conditions, under these redox potentials the species are controlled by the redox reactions of Mn and Fe [11], ORP-Eh between +250 and +100 mV, the metal oxyhydroxides are unstable and the process of dissolution of most metal oxides begins, and ORP-Eh values lower than −120 mV, the sediment is considered anoxic [11, 12]. If the oxidized layer is less than 5 cm, it is considered that the sediment is dominated by Mn^{2+} flow, the difference with the standard value of reduction of Mn (+526 mV), it is because this value is referenced to a pH = 7.0 and T = 25°C, while a value of +414 mV is more appropriate under environmental conditions.

2.2. Release of heavy metals and water contamination

In natural systems, the release and mobilization of metals normally occurs through soluble phases; so, the evaluation of the contamination of a soil or sediment cannot be based on its total concentration of the metal, since the potentially contaminating from metals depends on the chemical form or geochemical fraction in which the metal is found. In addition, the dissolved metal cations are subject to several mobilization or fixation processes depending on factors such as pH, ORP-Eh, presence of both soluble and insoluble organic matter, and ionic strength [13, 14]. The release, mobility and toxicity of metals depends both on the proportions of them in the geochemical fractions and the environmental conditions at the bottom of a body of water (pH, ORP-Eh, DO, OM and EC), [15–20]. However, several authors have shown the co-precipitation of some metals in the form of complexes with solid fractions [21].

The present metals in the sediments of a reservoir can eventually contaminate the water, when the physicochemical conditions of the hypolimnion and the benthic zone can produce processes such as desorption, dissolution of carbonates and reductive dissolution of oxides (Especially oxyhydroxides of Mn and Fe), mobilizing and releasing these metals, and other adsorbed: As and Cd [22]. In addition, the Mn and Fe can be mobilized as a product of the dissolution of sulfides generating H_2S, a product of the reduction of mineral sulfides such as Pyrite FeS_2 [23], or can be immobilized either by adsorption to volatile sulfides in acids (AVS: soluble FeS + insoluble FeS) under conditions of sulfate reduction [24], However, compared to Fe, adsorption of Mn is usually much lower in AVS [25]. In the environmental studies on metals in bodies of water, the topics of [26] are very important: the determination of the distribution and concentration of elements of environmental significance in the study area, the understanding the mobility of the elements through the different environmental systems and the identification of the relationships between the geochemical behavior of these elements and their impacts on biota.

Sediments act as sinks and sources of pollutants in aquatic systems due to their variable physical and chemical properties [27]. An important aspect when using soils and sediments as a record or indicator of metal contamination is to define the degree of enrichment, both natural and anthropic, according to the concentration of a contaminating metal. The sediment can be classified as contaminated or not, according to the proportion of metal concentration in relation to non-contaminated zones, according to its enrichment compared to threshold levels and according to sediment quality guidelines (SQC), [27, 28]. Among others, the Igeo Geoaccumulation Index, the FC Contamination Factor (FC) and the EF Enrichment Factor [29] are used. To determine the relative degree of contamination, comparisons are made with threshold concentrations of standard geological materials, such factors are commonly determined to characterize the magnitude of metal contamination in environmental samples [30]. However, one of the disadvantages of these systems of environmental soil evaluation is that they are based mainly on the total concentration and not on the geochemical fractions.

Contamination by heavy metals in the soil or sediment can represent risks for humans and the ecosystem through: direct ingestion or contact with contaminated soil, food chain (soil-plant-human or soil-plant-animal-human), consumption of contaminated soil water, decrease in the quality of food, reduction of land use capacity and problems of land tenure [31]. A metal reaches a body of water in various forms: mineral particle with potential to precipitate depending on the size and hydrological dynamics of the body of water, soluble free or forming complexes and adsorbed by solid phases; the soluble form can be rapidly adsorbed by other solid phases (interchangeable fraction) and is carried to the sediments when the soluble and insoluble metallic mineral phases pass from an oxic environment to an anoxic environment (precipitation or sedimentation), remaining under a complex series of biogeochemical processes [10, 32].

In the sediment, oxides, silicates, carbonates and organic matter are deposited more or less quickly. The organic matter being initially attacked in the sediments by bacteria, oxidizing it in its order with O_2, NO^{3-}, MnO_2, Fe $(OH)_3$, SO_4^{2-} and finally CO_2, transforming it into oxic water bodies, whose change of redox conditions occurs very close to the water-sediment limit. That gradient is usually referred to redoxcline. The redox potential measured in sediments is basically influenced by the microbial activity in its process of degradation of organic matter, which in turn depends on the availability and concentration of oxidizing agents [33, 34, 35]. The microbial degradation of the organic matter in the sediments is the mechanism that causes the reducing conditions, when an oxidation of the organic debris takes place, generating in the interstitial waters and base of the water column a decrease of the ORP-Eh values, whose reducing conditions can release metals such as Fe and Mn, affecting the quality of water and its potential use [6, 36].

The distribution of metals in the geochemical fractions in sediments should not be interpreted as a historical record of sedimentation, due to redox remobilization processes and the action of organisms in the upper layers of the sediment. The reducing processes under the water-sediment limit produce and release chemical compounds such as CH_4, NH_4^+, H_2S, Fe^{2+} and Mn^{2+}, while the oxidation of the organic matter produces HCO^{3-} and NH^{4+}, in this sense the abrupt decrease of dissolved oxygen is generated not only by the oxidation of organic matter but also by the oxidation of CH_4, NH^{4+} and HS^- [37, 38]. The models of release and mobilization of metals in the redox limit establishes a sedimentation or continuous deposition of detritus and mineral compounds from a column of hydrogen peroxide, and a source of fresh oxidized material from the upper and outer part, when this particulate material sediment and find the redox limit (oxic-anoxic region), it is reduced to its divalent forms, thus producing a point source of soluble metal in the redox limit; as the soluble metals diffuse (turbulent diffusion or molecular diffusion), up and down its point source generates two concentration profiles: an upper profile belonging to the insoluble species, caused by the gravitational accumulation of oxygen-metal hydroxides (M^{2+}), which dissolve at the redox limit and a lower profile belonging to the soluble species caused by the reductive dissolution of the solid phases of the metal, and whose concentration is higher towards the redox limit, where the migration upwards it, produces oxidation and formation of new oxidized insoluble species (**Figure 1**).

There are two basic approaches for the determination of the distribution of metals in the sediments: the theoretical one based on thermodynamic calculations that has the problem of incomplete data, and the other one is the experimental that involves a separation of fractions using a procedure or extraction sequential scheme (ESS), such as Tesier, BCR or others, these methods allow identifying phases or forms to be measured from very available to very low available [39, 40].

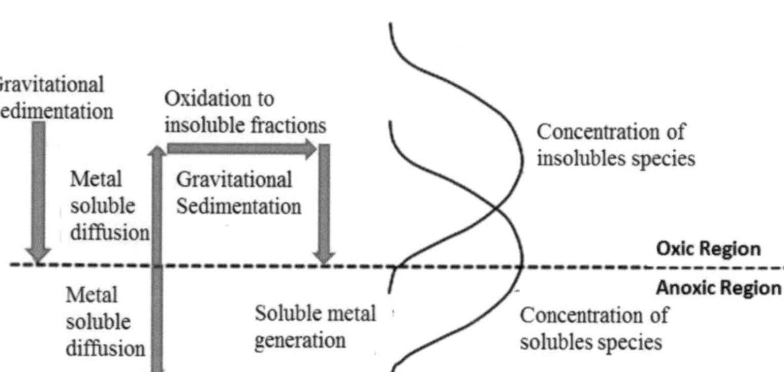

Figure 1. Conceptual model of the metals sensitive to redox processes (Mn and Fe), showing the profiles of the soluble and particulate forms. Source: adapted and modified [10].

3. Environmental study of water contamination and the advantages of the application of the geochemical fractionation methodology

3.1. Geochemical fractionation methodology

The geochemical fractionation is a laboratory process based on the application of selective chemical extractions, where the mobility of metals in soils and sediments is studied, imitating different environmental conditions or strong changes. This method or laboratory procedure is based on the rational use of a series of more or less selective reagents chosen to successively solubilize different mineralogical fractions that are believed to be responsible for retaining a large part of the trace elements [41]. Chemical extractions are used to evaluate operationally defined metal fractions, which may be related to particular chemical species, as well as to the potential mobility, bioavailability and ecotoxic phases of a sample, however, the results obtained by the extractions are dependent on the procedure applied [39, 42], so it is considered a very accurate approach to the environmental analysis of metal pollution in soils, sludges and sediments. Metals of anthropogenic origin are generally introduced into the environment as aqueous ions and inorganic complexes, which are easily adsorbed on suspended particles through weak chemical bonds to mineral grains and particles; these metals are predominantly belonging to the labile extractable fraction of the sediments [40].

In uncontaminated soils and sediments, trace metals occur mainly as relatively immobile species in silicates and primary minerals, as a result of weathering, a trace element fraction is gradually transferred to accessible forms for plants. In contaminated soils, the entrance of metals in almost all cases occurs in forms not associated with silicates. In the sediments, the situation is very similar, metallic species can exist in different forms [42]: in solution, ionic or colloidal, inorganic and organic complexes interchangeable, complexes where metals are strongly bound, insoluble mineral and organic phases and resistant secondary minerals. From a practical point of view, the geochemical fractionation of metals is achieved by sequential treatments that involve the use of chemical reagents, which are applied sequentially to a portion of the solid sample. The general applications of chemical extraction schemes are summarized in: characterization of pollution sources, evaluation of metal mobility and bioavailability and identification of metal link sites to evaluate the accumulation of metals, pollution and transport mechanisms.

In all the geochemical fractionation schemes, the extractants are applied in order to increase the reactivity to the successive fractions corresponding to forms with decreasing mobilities, most of the schemes or extraction sequences recommend the decomposition of the organic matter before the liberation of the subsequent metallic fractions [43]. Although sequential extractions are not specific to extracting the bound element from the solid fraction, they provide comparative information on metal mobility under changing environmental conditions, thus, the use of this approach can help to understand the relative contribution of metal sources and help in the prediction of trace element mobility [44]. Several extraction schemes have been designed for the determination of the forms of metals in soils and sediments, although with their limitations, sequential extraction is widely used to assess the relative importance of the different chemical forms that may be present in soils and sediments. However, the reagents used, the times and fractions extracted depend on the objectives set out in the research projects. The geochemical fractionation protocols present a common methodology based on the successive exposure of a solid sample to extractant solutions of increasing strength and/or of different characteristics. The chemical reagents are classified according to their mode of action [45]: concentrated inert electrolytes, weak acid, reducing agents, complexing agents, oxidizing agents and strong acids.

The sequential extraction schemes seek to solubilize the following fractions (**Figure 2**):

Exchangeable fraction: the metals in this fraction are bound to sediments or soil by weak adsorption in the particles, they are species bound by weak interactions of electrostatic type and released by ion exchange processes. Changes in the ionic strength of the interstitial water affects adsorption-desorption or ion exchange processes result in the release of metals at the sediment-water, soil-water interface.

Fraction bound to carbonates: metals bound to carbonates are sensitive to changes in pH, the increase in acidity liberates metal cations, achieving the release of the metal through the dissolution of a fraction of solid material. The metal fraction recovered in the acid soluble phase is co-precipitated with carbonates, and specifically adsorbed at some sites on the surface of the clays, organic matter and oxyhydroxides of Fe-Mn.

Fraction linked to oxides of Fe-Mn or reducible fraction: metals bound to oxides are unstable under reducing conditions, these conditions result in the release of metal ions. The oxides of Mn and Fe are excellent "sweepers" of metals.

Fraction linked to organic matter or oxidizable fraction: the degradation of organic matter under oxidizing conditions can release soluble metals bound to these materials. Trace elements can be incorporated in many forms to organic matter, including living organisms, organic layers on inorganic particles and biological detritus. In sediments and soils, the organic content comprises humic substances and to a lesser degree proteins, carbohydrates, peptides, amino acids and resins, whose organic material tends to be degraded under oxidizing conditions, allowing the release of the sorbed metals.

Residual fraction: this fraction contains minerals, which may contain metals in their crystalline networks. Primary and secondary minerals contain metals in their crystalline structures, constituting the total of this fraction.

In general, the exchangeable fraction corresponds to the way in which the metal is more available for consumption by the plants and that can be released with simply changes in the ionic strength of the medium [46]. The content of metals bound to carbonates is sensitive to changes in pH and becomes mobile when the pH decreases. The fraction of metals bound to oxyhydroxides can be mobilized with the increase or reduction of oxidizing conditions in the environment; finally, the fraction associated with the residual fraction can be mobilized only as a result of long-term weathering [43]. The different schemes for the extraction of metals during geochemical fractionation are based on sequences ranging from three to five or more steps.

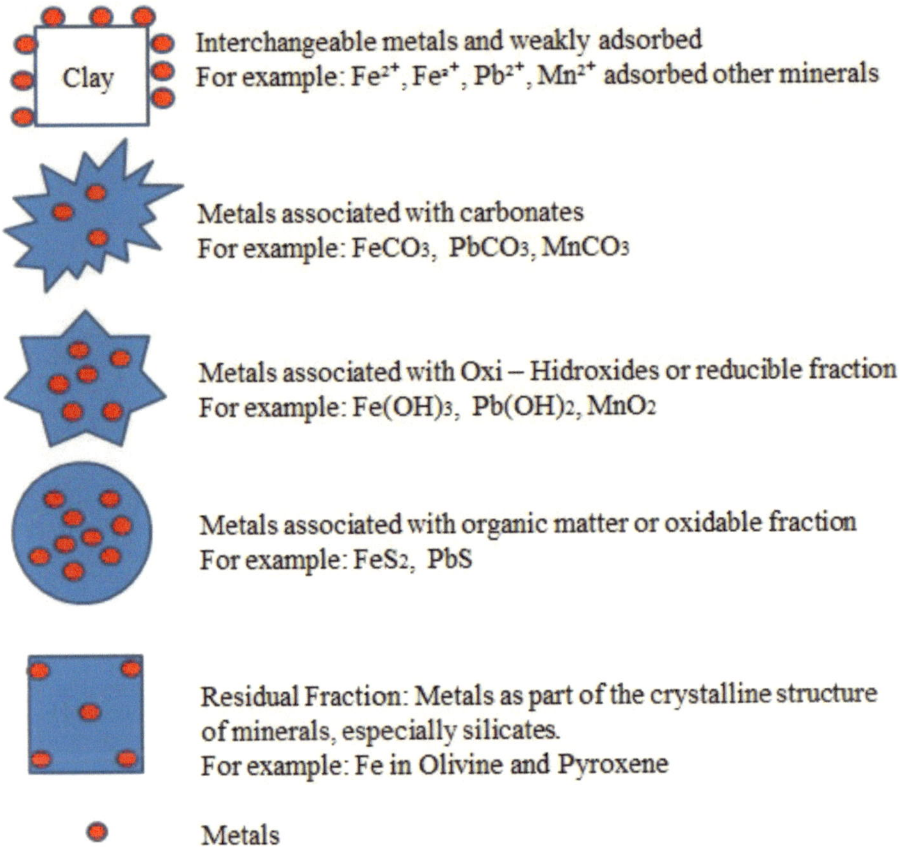

Figure 2. Metals forms or geochemical fractions of metals in sediments, soils or sludges.

3.2. The role of geochemical fractionation and the BCR methodology

The importance of geochemical fractionation or metals forms is not only to determine the groups in which metals are distributed in a soil or sediment, the most important thing is to be able to relate these results to the chemical conditions of the bottom water of a reservoir and the water of pore, and its variations in time (pH, EC, ORP-Eh and organic matter), and thus be able to infer when or under what conditions can begin to release metals. Among the established and well-known extraction schemes for metal extraction are:

3.2.1. Methodology of Tessier

It corresponds to one of the initial methodologies to establish the fractions of metals, developed in the year 1979 [47]. It determines five fractions of metals in matrices such as soils and sediments: Fraction 1: soluble in water and interchangeable, Fraction 2: associated with carbonates, Fraction 3: linked to oxides of Fe and Mn, Fraction 4: linked to organic matter and Fraction 5: residual, linked to complex oxides and silicates.

3.2.2. Methodology of bloom (emphasis on the extraction of non-volatile mercury fractions)

This methodology is very applied to mining and industrial sludge where Hg is one of the main pollutants, also determines five fractions, but with some differences: F1: soluble in water (Hg-w), F2: interchangeable (soluble in stomach acid Hg-I), F3: chelated organ (Hg-O), F4: elemental mercury (Hg-E) and F5: mercury sulfide (Hg-s) [48].

3.2.3. BCR methodology

Developed by the European Community, with the Measurement and Testing Program (BCR) [44], where several authors have defined the following fractions as the most important [40]. The BCR geochemical fractionation (European Community Reference Office—BCR), in order to determine the concentrations of metals according to their availability: BCR1: interchangeable metal, adsorbed and bound to carbonates, BCR2: metal reducible or associated with oxides, BCR3: metal associated with organic matter and sulfides and residual: mineral in silicates and other mineral phases [49].

The applied protocol in BCR Methodology [50] is:

- **Interchangeable fraction and soluble in acid, BCR1**: to 1 g of dry sediment is added 40 ml of 0.11 M acetic acid, it is stirred at 400 rpm, for 16 h. Centrifuge for 20 min at 3600 rpm, collect the supernatant for reading in AA or by ICP. The sediment is then washed with distilled water, stirring the sample with 20 ml, centrifuging 20 min at 3600 rpm, discarding the supernatant.
- **Reducible fraction, BCR2**: to the remaining sediment of the BCR1 extraction, 40 ml of 0.5 M hydroxylamine (pH = 1.5) is added, it is stirred at 400 rpm, for 16 h. It is then centrifuged 20 min at 3600 rpm, the supernatant is collected for reading in AA or by ICP. The sediment is washed with distilled water, stirring the sample with 20 ml, centrifuging 20 min at 3600 rpm, discarding supernatant.
- **Oxidizable fraction, BCR3**: to the remaining sediment of the BCR2 extraction, initially 10 ml of 8.8 M hydrogen peroxide is added, it is heated in water to 85° C for 1 h, and it is left to cool. The operation is repeated until all the organic matter is oxidized. About 50 ml of 1 M ammonium acetate is added at pH = 2.0, stirring at 400 rpm, for 16 h. Centrifuge for 20 min at 3600 rpm, collect the supernatant for reading in AA or by ICP. The sediment is washed with distilled water, stirring the sample with 20 ml, centrifuging 20 min at 3600 rpm, discarding the supernatant.
- **Residual metal**: the metallic fraction associated with the sediment after washing is considered the residual extracted with the mixture of hydrochloric acid (HCl) and nitric acid (HNO_3) in a ratio of 1:3.

The two reasons why the latter methodology was selected for the study of the mobility of Mn in the Riogrande II reservoir were the simplicity in the fractions with respect to the Tessier methodology and the existence of reference materials.

3.3. The complete environmental study for the metals in water reservoir

When the potential for water contamination is to be evaluated, based on the geochemical fractionation data of the sediments, the following analyzes are also included for the complete environmental study for the metals in water reservoir:

- Spatial time variations of the factors: pH, redox potential, electric conductivity, dissolved oxygen and temperature both in the bottom water and in the interstitial water.
- Proportions of organic matter, texture and mineralogy of the sediments using various techniques to associate mineral phases with the metals found in the fractionation.
- Experimental tests in columns with the sediments and the background water, making them vary in time especially in the pH and dissolved oxygen factors (modifying them), and evaluating the behavior of the metals in the water column.
- Analysis of soluble metals at least 50 cm above the water sediment limit and 1 m below that limit considering sampling points at least every 2 cm.
- Analysis of diffractometry with scanning electron microscopy, in order to verify the mineral phases predicted in the geochemical fractionation.

However, since a few years ago, the following three analyzes are being emphasized to assess the contamination of a metal in a soil, sediment or body of water: available fraction (corresponds especially to the BCR1 + BCR2 fractions, the extraction of the available phase is a simple extraction and can be done with: HCL 1 N-CH_3COOH 0.11 M-Na_2EDTA 0.05 M-$CaCl_2$ 0.01 M-NH_4NO_3 0.1 M-Ca $(NO_3)_2$ 0.1 M [28, 51, 52]), bioaccessible fraction (corresponds to the fraction of the substance that once released enters an organism via ingestion, inhalation or dermal contact [53, 54], the extraction of this fraction is also simple and is generally done with: acidified 0.4 M glycine (pH = 1.5) and it is then neutralized (pH = 7.0) that would correspond to the gastric and intestinal fluid [55, 56]) and bioavailable fraction (fraction that is accumulated in vegetable and/or animal tissues [51, 52]).

4. Case study: application of geochemical fractionation in the Colombian reservoir

4.1. The Riogrande II reservoir (Antioquia-Colombia), its location and other related information

In the Riogrande II reservoir (Antioquia-Colombia) there have been concentrations of Mn that eventually affected negatively the quality of the water resource, especially due to color effects, prior to the treatment in the potabilization plant that supplies water to the city of Medellin (Colombia). In order to implement the necessary controls, it was necessary to know and analyze the geochemical dynamics of Mn from the geochemical fractionation, its relations with the physicochemical conditions of the reservoir and finally determine the conditions of release from the sediment. The Riogrande II reservoir (6° 32.62 N, 75° 27.27 W) is located in the jurisdiction of the municipalities of Don Matías, San Pedro, Belmira, Santa Rosa de Osos and Entrerríos to the North of the city of Medellin (Department of Antioquia—Colombia) [57].

4.2. Mn in the sediments of this reservoir and its significance

In natural and artificial water bodies, such as reservoirs, sediments are important in water quality, since changes in redox conditions can dissolve oxy- hydroxides of Mn and Fe, mobilizing and releasing these metals [22]. The manganese is the third element of the most abundant transition metals in the earth's crust [58], as a transition metal it gives an important behavior at the level of oxide-reduction, and it facilitates the formation of complexes, being one of the most abundant elements of the earth's crust. It is found in the ground, sediments, rocks, water and in biological products; at least a hundred minerals contain it [59]. Mn is a metal that is very sensitive to redox changes in soils and natural waters. In humid soil conditions, Mn tends to be in its reduced and mobile form. It is well-known that the concentrations of soluble forms of Mn undergo strong changes in the depth of water columns where oxygen depletion occurs, which is strongly associated to the oxidation of organic matter in sediments; therefore, organic matter plays a very important role in the control and distribution of Mn^{2+} in the interstitial water; either, as an adsorbent or as a complexing ligand [60–62].

The behavior of Mn is influenced by the acidification of the medium, with an inverse correlation between the concentrations of Mn^{2+} and the pH values. Some bodies of water with pH values ≥6.0, act as sinks of Mn, while smaller values act as a source of Mn. This behavior is possibly controlled by dissolution processes of species such as Mn^{4+} and Mn^{3+}, the oxidation kinetics of Mn and the equilibrium of ion exchange [63]. Both the humic fraction and the concentrations of metal adsorbed to insoluble phases decrease as the pH decreases [64, 65]. The Mn^{2+} ion released at the bottom, forms organic complexes or the hexacuous ion Mn $(H_2O)_6^{2+}$ that is very stable. When the redox conditions are adequate, a precipitate of MnO_2 (Pyrolusite) is formed; it forms incrustations even at concentrations of 0.02 mg L^{-1}. The environmental problem of Mn in reservoirs has been correlated with thermal and chemical stratification, since some bacteria use Mn^{4+} as an oxidizing agent of organic matter in its metabolic processes, reducing it to Mn^{2+} and releasing it to the water column. It has even been considered that oxide-reducing bacteria of Fe also act on manganese [66].

4.3. Application of the BCR methodology, results, discussion and conclusions

The application of BCR geochemical fractioning of Mn in the Riogrande II reservoir is presented, starting from the design of research, applied methodology, results, discussion and analysis, finally arriving at the proposed model of Mn entrance to the reservoir, the distribution of its forms in the sediment and release mechanisms towards the water column.

4.3.1. *Used methodology*

A sediment sampling was carried out in two different months (March and June 2008), in seven stations within the reservoir, using the Ekkman dredger and through AA, the total Mn was determined. For the geochemical fractions, the BCR methodology was applied. The sampled stations were characterized by: E1. Rio Chico entrance: station located at the entrance of the Chico River in the reservoir, in the extreme west, depth of sampling 13 m., E2. Middle Riochico: located in the middle part of the Chico River, depth of sampling 27 m., E3. Rio Grande entrance: located at the entrance of Rio Grande, depth of sampling 12 m., E4. Middle Rio Grande: located in the middle part of Rio Grande, similar to the E2 by the amplitude of the reservoir, depth of sampling 35 m., E5. Las Animas river mouth: located at the entrance of Las Animas Creek, depth of sampling 11 m., E6. Intake Tower: located in the reservoir dam site, its importance is that in this site the water is captured by the treatment plant, depth 18 m. and E7. Dam Zone: it was located in the dam area, depth of sampling 42 m. After taking the sediment, the analyzes were carried out in situ by introducing the electrodes into the sediment-water mixture of factors such as: pH, dissolved oxygen, redox potential and electrical conductivity. After reading the interstitial water variables, the sediment-water mixtures of each sampled station were packed in vacuum polypropylene bags and immediately preserved in cold temperature until their arrival to the laboratory, where they were stored in the refrigerator until further analysis of Mn [67].

The analyzed factors for the water matrix (bottom hypolimnion and interstitial) were pH, dissolved oxygen, ORP-Eh potential and electrical conductivity, and in the sediment the percentage of organic matter (%OM). For the dependent variables in the sediment, were analyzed: total Mn, the Mn fractions according to the BCR protocol (interchangeable Mn and associated with carbonates, Mn associated with oxides, Mn associated with organic matter and sulfides), and the residual fraction (Mn associated with silicates). Finally, the identification of the possible mineral phases associated with Mn in some of the sediment samples was carried out by means of the analysis of X-ray diffractometry coupled to scanning electron microscopy (XRD-SEM). To the water samples of the bottom hypolimnion and interstitial water were measured pH and temperature with the WTW-model 330 equipment with a glass electrode, the electrical conductivity with the WTW-model 720 conductivity meter, the ORP-Eh potential with the WTW-model 330 equipment with a platinum electrode and dissolved oxygen with WTW brand oximeter with Cellox® cell. The organic matter (% OM) of the sediment was determined in the same way as for the soil, through the volumetric method of Black and Wlakley.

For total Mn and BCR fractions, the sediment brought from the reservoir was dried at environmental conditions for 1 week; for the total Mn, to 1 g of the dry sediment, a mixture of hydrochloric acid (HCl) and nitric acid (HNO_3) was added in a 1:3 ratio for the extraction, following the ISO 11466 protocol. The interchangeable fraction and soluble in carbonate (Mn) -BCR1) was obtained by stirring 1 g of dry sediment for 16 h with 40 ml of 0.11 M acetic acid at 400 rpm and then centrifuged, and then the supernatant was filtered, the reducible fraction (Mn-BCR2) was obtained by stirring the previous residual sediment for 16 h with 40 ml of 0.5 M hydroxylamine chloride, acidified to pH 1.5 with nitric acid, stirred at 400 rpm and subsequent centrifugation, extraction and filtration of the supernatant, for the oxidable fraction (Mn-BCR3) Initially double oxidation was performed with 20 ml of 0.88 M hydrogen peroxide for 2 h, and then agitation of the remaining sediment with 40 ml of 1 M ammonium acetate for 16 h and centrifugation, extraction and filtration of the supernatant, and finally for the remaining residual sediment fraction, it was treated with a mixture of hydrochloric acid (HCl) and nitric acid (HNO_3) in a ratio of 1:3.

The readings for the total Mn as well as for the fractions were carried out by Atomic Absorption in a GBC Plus equipment, the detection limit (LD) was 0.01 mg L^{-1} and the limit of quantification was 0.05 mg L^{-1}. All analyzes were performed by duplicate. As a decision rule, the efficiency of the geochemical fractionation for each sediment sample was calculated by using the percentage of recovery (% Recovery), which relates the total metal and the resultant of the sum of the fractions BCR1, BCR2, BCR3 and residual (Res), the optimal condition establishes that there should not be more than ±10% in the difference of both concentrations (100 ± 10%, [68]), this range is considered a quality standard of the extraction process.

The averages, standard deviations and the level of variability were determined by the coefficient of variation, both for the total Mn dependent variables and the Mn fractions (BCR1, BCR2, BCR3 and residual) and for the independent variables: pH, redox potential, electrical conductivity, dissolved oxygen and organic matter. To evaluate the hypothesis, the analysis of components of variance was initially started in order to determine the factors (sampling time and station) that contributed to the Mn (total and fractions), afterwards a Pearson multiple correlation was made ($p < 0.05$), to establish the physicochemical factors that have a statistically significant correlation with the response variables, moving on to an analysis of variance between the values of Mn (total, BCR1 and BCR2) and the factors: sampling, station and the physicochemical conditions of the bottom hypolimnion water and sediment ($p < 0.05$). Before proceeding with the calculations, the assumptions of the parametric statistics were corroborated, for this vector was evaluated in the matrix of the following dependent variables (Mn-Total, Mn-BCR1 and Mn-BCR2), normality was evaluated with the Shapiro Wilks (SW) Test and homoscedasticity with Bartlett's Contrast Value (β). Statistical calculations were performed in the Statgraphics Centurion XVI program with significance 0.05.

4.3.2. *Results of the geochemical fractionation*

The results of the total Mn (mg kg^{-1}) are summarized in **Table 1**.

Value (mg kg^{-1})		Location	Sampling period
Max	651,25	E1	Rainy (March 2008)
Min.	121,17	E2	
Average	407,04 ± 194,7 VC = 47,83 %	For the 7 stations	
Max	520,50	E3	Dry (June 2008)
Min	183,50	E2	
Average	334,50 ± 122,1 VC = 36,50%	For the 7 stations	
Variation between both sampling periods	In the dry period, the concentration of Total Mn was reduced by 17.82% with respect to the rainy period. However according to the CV, the difference is not significant.		

CV: coefficient of variation

Table 1. Synthesis of results of the Total Mn for the geochemical fractionation.

For both samplings, the results showed the highest concentrations in the two main entrances to the reservoir: E1 and E3, while in the middle of the Riochico, it had the lowest concentrations; when comparing these values against the considered threshold value (400 mg kg^{-1} [69]), it was found that stations E1, E3 and E4 (Middle of Rio Grande), had at least the threshold value. There were practically no significant differences between the two sampling periods. At the level of the BCR1 geochemical fraction, the results were summarized in **Table 2**.

Value (mg kg^{-1})		Location	Sampling period
Max	336,11	E1	Rainy (March 2008)
Min	109,78	E2	
Average	236,72 ± 79,9 VC = 33,77 %	For the 7 stations	
Max	335,94	E3	Dry (June 2008)
Min	100,00	E7	
Average	226,4 ± 91,7 VC = 40,5 %	For the 7 stations	
Variation between both sampling periods	In the dry period, the concentration of Mn-BCR1 was reduced by 4.36% compared to the rainy period. However according to the CV, the difference is not significant		

CV: coefficient of variation

Table 2. Synthesis of Mn-BCR1 results.

Similar to the Total Mn, the highest concentrations were found at the entrances to the reservoir (E1 and E3), in none of the stations and climatic periods, the threshold of 400 mg kg^{-1} was exceeded. There were practically no significant differences between the two sampling periods. At the level of the BCR2 geochemical fraction, the results were summarized in **Table 3**.

Value (mg kg^{-1})		Location	Sampling period
Max	79,86	E1	Rainy (March 2008)
Min	0,79	E7	
Average	30,17 ± 29,5 VC = 97,68 %	For the 7 stations	
Max	68,74	E1	Dry (June 2008)
Min	12,20	E2	
Average	42,8 ± 19,5 VC = 45,5 %	For the 7 stations	
Variation between both sampling periods	In the dry period the concentration of Mn-BCR2 was increased by 41.86% with respect to the rainy period. However according to your CV, the difference is not significant		

CV: coefficient of variation

Table 3. Synthesis of Mn-BCR2 results.

The reducible geochemical fraction Mn-BCR2 did not exceed 80 mg kg^{-1}, unlike the total Mn and Mn-BCR1, this fraction increased in the dry period with respect to the rainy period, but not significantly. The whole analysis of the shown fractions as a proportion, indicated a predominance of Mn in the BCR1 and BCR2 fractions, with respect to the Mn of BCR3 and residual; however, comparing both sampling times, the proportions in the dry period in stations E1, E2, E3, E4, E5 and E6 (Rio Chico, Rio Grande and Las Animas, Intake tower), exceeded the concentrations of the rainy period (**Figure 3**).

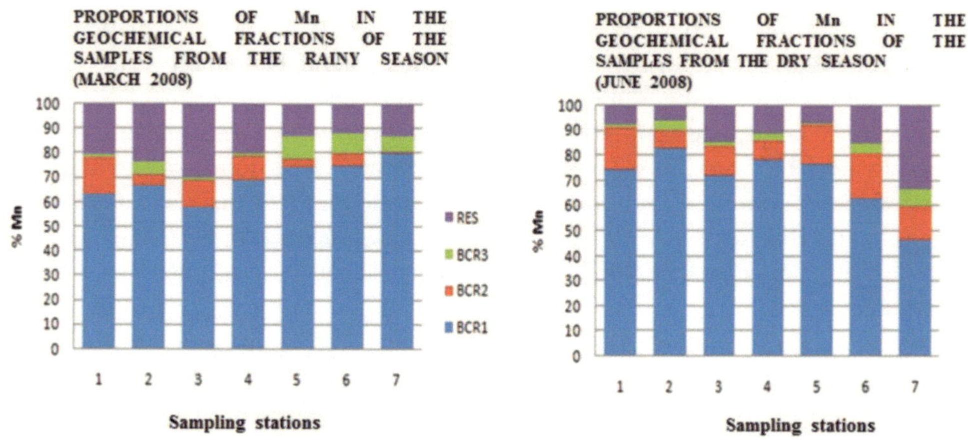

Figure 3. Proportions of the fractions of Mn in two sampling periods in seven stations. Note that the available fractions BCR1 Mn and BCR2 Mn, exceed 60% of the total Mn.

BCR1: co-precipitated, interchangeable and associated with carbonates Mn.

BCR2: reducible (associated with oxyhydroxides) Mn.

BCR3: oxidizable (Associated with sulfides and insoluble organic matter) Mn.

Res: associated with the residual fraction (silicates) Mn.

Finally, establishing the proportion of available Mn as the ratio between the sum of the BCR1 and BCR2 fractions and the Total Mn. It was found that more than 80% of the Mn of the sediment was available in the stations: E1 (Rio Chico entrance), E4 (Middle of Rio Grande), E5 (Las Animas) and E6 (Intake tower) for the dry period and E2 (Middle of Rio Chico) for the rainy period. The recovery percentage for the fractionation produced values between 80 and 90%, and indicated that the sequential extraction scheme could not solubilize the whole Mn, it was probably due to the formation of very stable organo-metallic compounds, especially in the samples taken during the rainy period. Likewise, the E2 sample (Middle of Rio Chico) presented a recovery of 135%, and could be associated with contamination during the extraction process.

4.3.3. Results-physicochemical analysis of sediment and bottom water

The results of the physicochemical analysis of two types of water in the bottom of the reservoir were compared: bottom hypolimnion water and interstitial waters (sediment water and pore water). This comparison was carried out with the purpose of identifying the changes in geochemical conditions between the water on the sediment (bottom hypolimnion) and the water under the water-sediment limit (interstitial water) and its relation with the results of total Mn and Mn in the fractions.

The results indicated an oxic bottom hypolimnion and sub-toxic interstitial water (sediment water) during the rainy period, the redox potential was more variable in the waters of the bottom hypolimnion where positive and negative redox potentials were present, compared to the same potential in interstitial waters, where the records always showed negative potentials (−128.4 mV in the rainy period and −120.3 mV in the dry period); for the sediment, the organic matter had values between 25.01% in the rainy period and 24.5% in the dry period, neutral pH in the water of the bottom hypolimnion, and more acid for the interstitial waters. The fractionation showed that more than 50% of the Mn in the sediments is present in available form (69% in the rainy period and 79% in the dry period), probably co-precipitated and adsorbed in the organic matter.

pH: the pH of both types of water (bottom hypolimnion and interstitial), presents values close to neutrality; however, while the waters of the bottom hypolimnion is practically neutral; the interstitial waters were more acidic in the rainy period.

Redox potential-Eh: the waters of the bottom hypolimnion changed significantly from positive redox potentials in the rainy period to negative values in the dry period, but always remained under reducing conditions with a maximum of +82.5 mV. In general, the water of the bottom hypolimnion of Riogrande II reservoir can be considered as a reducing agent, despite having positive redox potentials, and the interstitial or sediment waters can be considered as strongly reducing agent.

Dissolved oxygen (DO): in terms of dissolved oxygen, the waters of the bottom hypolimnion, maintained both oxic and hypoxic concentrations, while the sediment waters remained hypoxic. The dissolved oxygen presented a different behavior between the waters of the bottom hypolimnion and the interstitial waters. The decrease in dissolved oxygen, both in the water of the bottom hypolimnion and in the interstitial water for the dry period could be associated with a higher concentration of soluble organic matter.

Electrical conductivity (EC): first, when analyzing the values, it is found that the electrical conductivity is almost 9 times higher than the values of the hypolimnion of the bottom with respect to the interstitial waters; then, the conductivity values do not show significant variability depending on the time of sampling.

Organic matter (% OM): in the month of the rainy period, the average value in the stations was 25.01 ± 3.9% of organic matter, with a CV of 3.9%, the highest proportion was found in the sediment of E2 with 32.28% and the minimum in E5 with 21.01%; for the dry period, the value was similar with 24.05 ± 2.9%, and a CV of 12%. From the rainy period to the dry period, it was increased the % of organic matter, at stations E4 (Middle of Rio Grande), E6 (Intake Tower) and E7 (Dam Zone).

4.3.4. Results of the EDX-SEM analysis

Sediment samples taken during the rainy period were analyzed; the results showed that in the E2 station the Mn is associated with particles of Pb and calcium carbonate (**Figure 4**), in station E4 there were Zn sulfide (Characteristic of reducing conditions) and in the associated diffractogram, besides the Zn, Mn is associated with Fe within a matrix that is composed of Si, Al, C and O, probably associated with aluminosilicates and carbonates; in E6 the sample did not present evidence of Mn, only Ti within a matrix of aluminosilicates. Finally, the E7 sample shows some mineral phases associated with Mn carbonates, free spherical forms (less than 1 μm), and carbonates associated with Pb that are abundant in the sediments of this station (**Figure 5**).

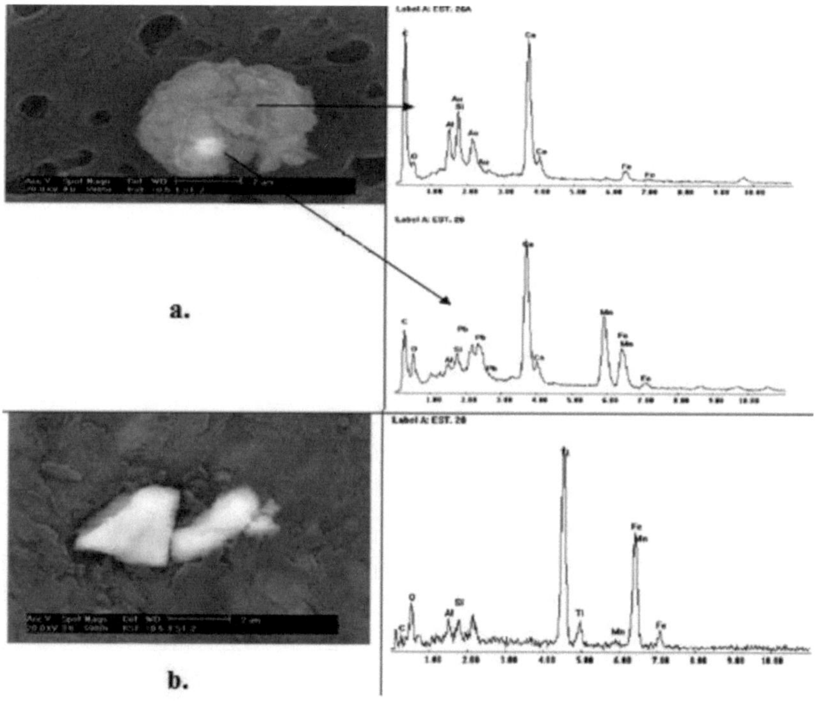

Figure 4. Image and diffractograms of the sediment in E2, observed in (a), a clear point corresponding to Pb associated with Mn and calcium carbonate, the gray part of the particle is mainly calcium carbonate. In the image (b) there are particles containing Mn associated with Ti.

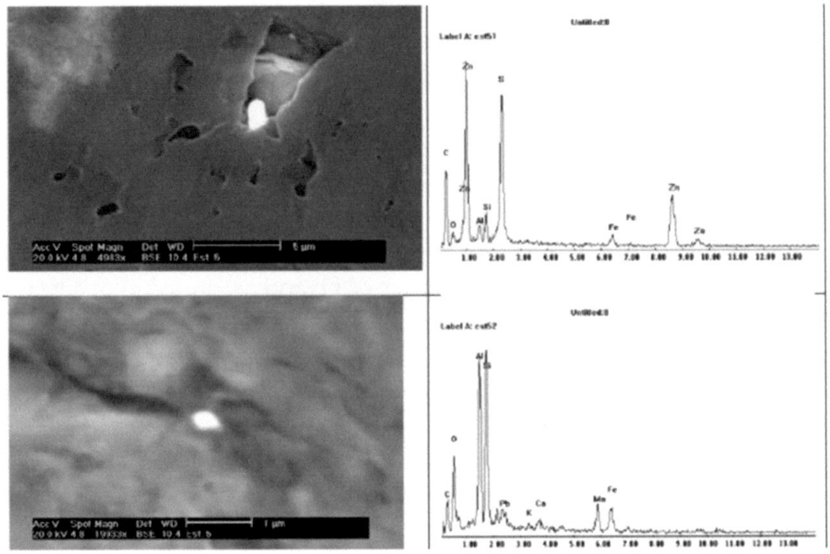

Figure 5. Images and diffractograms of the sample in E7, the particle in the image above corresponds to Pb carbonate and Mn carbonate, the diffractograms showed signal for Mn.

4.3.5. Discussion and analysis of results

- Of the geochemical fractionation and its relation with the physicochemical conditions of the bottom water and interstitial water

The entries of the Chico and Rio Grande Rivers (E1 and E3), correspond to the tributaries in which the highest concentrations of Mn reached the sediment of the reservoir, and although statistically it was not significant, there was a greater contribution at the time of rains, possibly due to an increase in the soluble and bottom charge of both streams. Conditions associated with the dry weather period could be the reason that the available fractions of Mn (MnBCR1—Interchangeable Metal and Carbonate Associated and MnBCR2—Oxyhydroxides) have been higher compared to the rainy period, possibly associated with the decrease in the redox potential during the dry period, especially in the water of the bottom hypolimnion, and could be associated with the decrease in the flow rate and a decrease in dissolved oxygen, however statistically it was not significant. The low values of redox potential, both in the bottom hypolimnion (less than 100 mV) and in the sediments (up to −150 mV), shows that in this anoxic-reducing environment, it would be changing from a post-oxic zone in the bottom hypolimnion to a sulfidic zone in the sediments [60], where Mn^{2+}, Fe^{2+} and NH^{4+} occur due to the presence of Zn sulfide in the sediments; the probability of finding oxides of Mn in this environment is minimal and the Mn found is in a reduced form, complexed and adsorbed to other mineral phases. However, no pyrite was found in the EDX-SEM analyzes, which could mean that the presence of HS- and H_2S necessary for the formation of this mineral, is below 10 cm.

- Possible causes of the release of Mn and its remobilization to the water column

One of the possible causes of the release and remobilization of Mn in the sediments of the Riogrande II reservoir were the highly reducing conditions of the sediments generated by microbial oxidation and the decrease in pH. These conditions reduce the phases of oxidized metal, releasing Mn^{2+} among other ions, which are solubilized and complexed with soluble organic matter such as fulvic acids [59]. The strongly reducing conditions in the interstitial waters remained practically the same for both sampling periods (−120 mV), and as for the pH, it went from acidic conditions (5.4) in the rainy period to almost neutral (6.7) in the dry period; however for the bottom hypolimnion waters, if there was a significant reduction in the redox potential in the dry period, without significant changes in the pH, which could suggest that the change from reducing to strongly reducing conditions in the bottom hypolimnion facilitates the remobilization of metals such as Mn.

The release of metals from the sediment of the water column may be due to processes such as the desorption and formation of soluble organic complexes, associated with the decomposition of organic matter [70], processes such as dissolution reductive metal oxides are possibly the cause of the release and mobility of Mn in the water sediment interface of the Riogrande II reservoir; however, it should be considered that the process of reducing the oxides of Mn is at the highest level in the column of water over the limit water-sediment [71]. One of the characteristics of the sediment found in the reservoir is the continuous presence of sulfides as demonstrated by the EDX-SEM analysis, it indicates strongly reducing sediment conditions during sampling; however, the presence of some carbonates in stations such as E2 (Middle of Riochico), E3 (Rio Grande Entrance) and E7 (Dam area), added to the pH values in interstitial waters, especially in the dry period, allows us to conclude that the sediments in the Riogrande II reservoir are strongly of reducing characteristics with conditions from neutral to acidic. The fractionation results indicate that although the Mn would not be found in high levels of sediment contamination, there is an environmental risk for the water quality of the reservoir due to the high proportion (more than 50%) of this metal in the more mobile fractions. (BCR1 and BCR2), especially in the dry period for E1, E3 and E4, these fractions are highly bioavailable [72]. It can be established that the Mn that enters the Riogrande II reservoir, probably makes it both soluble (organic and inorganic complexes) and insoluble (Oxyhydroxides-aluminosilicates), and when precipitated towards the bottom of the water column.

- Proposed model of Mn remobilization

Despite the presence of significant concentrations of total Mn in the sediment (values between 121.17 and 651.25 mg kg^{-1}), only Mn carbonate evidence was found in the sediment in the prey area. The results of the geochemical fractionation show that this metal is mainly in adsorbed form to other minerals, followed by Mn associated with organic matter and some oxides and hydroxides of Mn (**Figure 6**).

From the geochemical fractionation, the EDX-SEM findings and complementary studies of the sediment water interface along with geochemical modeling allowed to identify the following

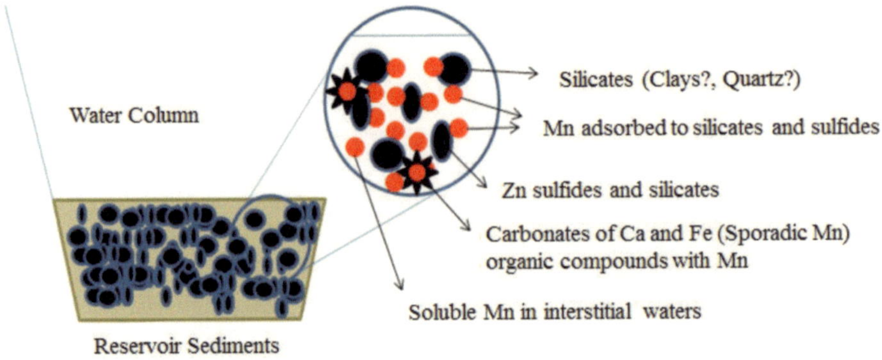

Figure 6. Probable way in which Mn^{2+} is found in the sediments of the Riogrande II reservoir in an adsorbed form, within the interstitial waters and sporadically as carbonates according to evidence from the geochemical fractionation and EDX-SEM analysis.

processes for the Mn in the Riogrande II reservoir: entrance to the reservoir in mineral form (silicates and oxides), and complex form, beginning of the reductive dissolution in terms that the oxides enter the water with reducing potentials, release in the hypolimnion of Mn^{2+} ions, adsorption of Mn^{2+} to other minerals and organic compounds, precipitation in the reservoir and finally formation of organic and inorganic complexes of Mn^{2+} in the same water column.

5. Conclusions

- The proportion of Mn available (BCR1 and BCR2) in the sediments was in a range above 50% of the total Mn in both climatic periods, which indicates that for the sediments of the Riogrande II reservoir, most of the Mn corresponds to fractions possibly adsorbed to mineral phases and organic compounds, where physical processes such as resuspensions generate chemical changes of pH, redox potential and conductivity, among others. It allows the transfer of insoluble phases with little mobility to mobile phases in the water column. This condition creates a danger to the water quality of the reservoir, because whenever there are falls of redox potential and/or the pH in the sediment water limit is lowered, reductive dissolution and desorption processes will be generated and will mobilize heavy metals towards the column of water.
- With the results and statistical analysis obtained, we accept the hypothesis about mobilization scenarios of the Mn present in the sediment to the water column, where the reducing conditions, acid pH and high conductivity increase the concentration of Mn available, product of processes of remobilization in the sediment.

Acknowledgements

The authors thank the investigation group "La Salada", Sennova strategy of the SENA (Colombia) and the Center of Renewable Natural Resources, La Salada (Antioquia) for their support in the edition of this chapter. And also the investigation group "GDCON" of the University of Antioquia (Colombia) for their support in the investigation.

Author details

Juan Pablo Salazar Giraldo[1,2*]

*Address all correspondence to: jpsg0501@gmail.com

1 Geologist, Caldas University, Colombia

2 Engineering, Antioquia University, Center of Renewable Natural Resources, "La Salada", SENA, Antioquia, Colombia

References

[1] Zhang C, Yu Z, Zeng G, Jiang M, Yang Z, Cui F, Zhu M, Shen L, Hu L. Effects of sediment geochemical properties on heavy metal bioavailability, review. Environment International. 2014;**73**:270-281

[2] Tang W, Duan S, Shan B, Zhang H, Zhang W, Zhao Y, Zhang C. Concentrations diffusive fluxes and toxicity of heavy metals in pore water of the Fuyang River, Haihe Basin. Ecotoxicology and Environmental Safety. 2016;**127**:80-86

[3] Betancourt C, Suarez R, y Toledo L. 2010. Variabilidad Iónica y características del Embalse de Abreus, Cuba. Limnetica 2010;**29**(2):341-352

[4] UNESCO, WHO, UNEP. In: Chapman D, editor. Water Quality Assessments: A Guide to the Use of Biota, Sediments and Water in Environmental Monitoring. 2nd ed. London: E & FN Spon; 1996

[5] Betancourt C, Suarez R, y Toledo L. Patrones de distribución temporal de algunas vari ables físicas y químicas en el embalse Paso Bonito, Cienfuegos – Cuba. Limnética 2009; **28**(1):23-34

[6] Salomons W, Baccini P. Chemical species and metal transport in lakes. In: Brinkman, Sadler, editors. The Importance of Chemical "Speciation" in Environmental Processes. 1986. pp. 193-216

[7] Garcia GLJ, y Camacho A. Anaerolimnologia: pequeña guía para el muestreo en aguas Microaerobicas y anoxicas en lagos y embalses estratificados. Limnetica 2001;**20**(1): 179-186

[8] Savenko AV, Savenko VS, Pokrovskii OS. Experimental simulation of chemistry transformation of anaerobic water during aeration. Water Resources. 2016;**43**(4):647-656

[9] Wetzel R. Limnology. 2nd ed. Vol. 1985. Saunders College Publishing CBS. p. 767

[10] Davison B. Iron and manganese in lakes. Earth Sciences Review. 1993;**34**:119-163

[11] Burton ED, Phillips IR, Hawker DW. Geochemical partitioning of copper, lead and zinc in benthic, estuarine sediment profiles. Environmental Quality. 2005;**34**:263-273

[12] Rozanov AG. Oxidation – Reducing processes in marine sediments and methods of their studies. In: Khimicheskiy Analiz Morskikh Osadkov (Chemical Analysis of Marine Sediments). Moscow: Nauska; 1988. pp. 5-44

[13] Traina SJ, Laperche V. Contaminant bioavailability in soils, sediments, and aquatic environments. Papers from a National Academy of Sciences Colloquium Paper USA. 1999;**96**:3365-3371

[14] Yang J, Chen L, Li-Zao L, Wei-Ling S, Xiang-Zhou M. Comprehensive risk assessment of heavy metals in lake sediment from public parks in shanghai. Ecotoxicology and Environmental Safety. 2014;**102**:129-135

[15] Hamdoun H, Van-Veen E, Basset B, Lemoine M, Coggan J, Leleyter L, Baraud F. Characterization of harbor sediments from the English Channel: Assessment of heavy metal enrichment, biological effect and mobility. Marine Pollution Bulletin. 2015;**90**: 273-280

[16] Azam M, Tripathi JK. Recent contributions in the field of sediment geochemistry. Revie Article. Proceedings of the Indian National Science Academy. 2016;**82**(3 July Spl Issue):805-815

[17] Inaotombi S, Gupta PK. Fate of metals in relation to water and sediment properties in a subtropical Lake in central Himalaya, India. Archives of Environmental Contamination and Toxicology. 2017;**72**(3):391-401

[18] IUPAC. Guidelines for terms related to chemical speciation and fractionation of elements. Definition of structural aspects and methodological approaches. Pure and Applied Chemical. 2000;**72**(8):90-110

[19] Dabrin A, Durand CL, Garric J, Geffard O, Ferrari BJD, Coquery M. Coupling geochemical and biological approaches to assess the availability of cadmium in freshwater sediment. The Science of the Total Environment. 2012;**424**:308-315

[20] De Jonge M, teuchies J, Meire P, Blust R, Bervoets L. The impact of increased oxygen conditions on metal contaminated sediments part I: Effects on redox status, sediment geochemistry and metal bioavailability. Water Research. 2012;**46**:2205-2214

[21] Schüring J, Schulz HD, Fischer WR, Böttcher J, Duijnisveld WHM, editors. Redox: Fundamentals, Processes and Applications. Springer-Verlag; 1999. p. 273

[22] Wildman JRA, Dalleska NF, Anderson M, Hering JG. Effect of changes in water level on sediment pore water redox geochemistry at a reservoir shoreline. Applied Geochemistry. 2010;**25**:1902-1911

[23] Delfino JJ, Lee FG. Variation of Manganese, dissolved oxygen and related chemical parameters in the bottom waters of Lake Mendota, Wisconsin. Engineering Experiment Station Reprint. 1971. no. 1551. pp. 1-11

[24] Mucci A, Boudreau B, Guinard C. Diagenetic mobility of trace elements in sediments covered by a flash flood deposit: Mn, Fe and as. Applied Geochemistry. 2003;**18**:1011-1026

[25] Koretsky C, Haas JR, Miller D, Ndenga N. Seasonal variations in pore water and sediment geochemistry of littoral Lake sediments (asylum lake, MI, USA). Geochemical Transactions. 2006;7(11):1-26

[26] Ashley R, Bailey E, Balistrieri L, Foster A, Gough L, Gray F, Koski R, Rytuba J, Seal R, Smith K, Stillings L, And Wantly R. Introduction to pathways of metal transfer from mineralized sources to bioreceptors. In Pathways of Metal Transfer from Mineralized Sources to Bioreceptors: A Synthesis of the Mineral Resources Program's Past Environmental Studies in the Western United States and Future Research Directions. U.S. Geological Survey Bulletin 2191 .2003

[27] Praveena SM, Ahmed A, Radojevic M, Abdullah MH, Aris AZ. Factor-cluster analysis and enrichment study of mangrove sediments-an example from Mengkabong, Sabah. The Malaysian Journal of Analytical Sciences. 2007;**11**(2):421-430

[28] Lesmes L. y Cedeño CJ. Un Factor de Movilidad en Geoquímica Ambiental. In II Simposio Internacional de Geoquímica Ambiental. Cartagena-Colombia. 2006. p. 196-200

[29] Thomilson DC, Wilson DJ, Harris CR, Jeffrey DW. Problem in heavy metals in estuaries and the formation of pollution index. Helgoländer Meeresuntersuchungen. 1980;**33**(1-4):566-575

[30] Kartal S, Aidin Z, Tokalioglu S. Fractionation of metals in street sediment samples by using the BCR sequential extraction procedure and multivariate statistical elucidation of the data. Journal of Hazardous Materials. 2006;**132**(1):80-89

[31] Wuana RA, Okieimen FE. Heavy metals in contaminated soils: A review of sources, chemistry, risks and best available strategies for remediation. Review Article. International Scholarly Research Network ISRN Ecology. Article ID 402647. 2011;**2011**:20

[32] Forstner U, Wittman GTW. Metal Pollution in the Aquatic Environment. 2nd Revised ed. Berlin, Heidelberg, Germany: Springer-Verlag; 1983. p. 485

[33] Hallberg RO, Larsson C. Biochelates as a cause of metal cycling across the redoxcline. Aquatic Geochemistry. 1999;**5**:269-280

[34] Cappellen PV, Wang Y. Cycling of iron and manganese in surface sediments: A general theory for the coupled transport and reaction of carbon, oxygen, nitrogen, sulfur, iron and manganese. American Journal of Science. 1996;**296**:197-243

[35] Park ASS, Jaff PR. Development of a sediment redox potential model for the assessment of postdepositional metal mobility. Ecological Modelling. 1996;**91**:169-181

[36] Tipping E. Cationic Binding by Humic Substances. Cambridge Environmental Chemistry sciences. Cambridge University Press; 2004. p. 446

[37] Furrer G, Werhli B. Biogeochemical process at the sediment-water interface: Measure ments and modeling. Applied Geochemistry (Supplement). 1993;**2**:117-119

[38] Boudreau BP. Metals and models: Diagenetic modelling in freshwater lacustrine sediments. D.G. Frey & E.S. Deevey review #2. Journal of Paleolimnology. 1999;**22**:227-251

[39] Rauret G, Lopez-Sanchez JF, Sauquillo A, Rubio R, Davison C, Ure A, Quevauviller P. Improvement of the BCR three step sequential extraction procedure prior to the certification of new sediment and soil reference materials. Journal of Environmental Monittoring. 1999;**1**:57-61

[40] Jain CK, Malik DS, Rashmi Y. Metal fractionation study on bed sediments of Lake Nainital, Uttaranchal, India. Environmental Monitoring and Assessment. 2007:129-139

[41] Gleyzes C, Tellier S, Astruc M. Fractionation studies of trace elements in contaminated soils and sediments: A review of sequential extraction procedures. Trends in Analytical Chemistry. 2002;**21**(6 and 7):451-466

[42] Hlavay J, Prohaska T, Weisz M, Wenzel WW, Stingeder J. Determination of trace elements bound to soils and sediment fractions. International union of pure and applied chemistry - IUPAC, analytical chemistry division. Pure and Applied Chemistry. 2004;**76**(2):415-442

[43] Filgueiras AV, Lavilla I, Bendicho C. Chemical sequential extraction for metal partitioning in environmental solid simples. Tutorial review. Journal of Environmental Monitoring. 2002;**4**:823-857

[44] Lopez-Sanchez JF, Pueyo M, Sastre J, Hernandez E, Vidal M, Reuret G. Heavy metals in environment. Prediction of trace element mobility in contaminated solis by sequential extraction. Journal of Environmental Quality. 2003;**32**:2054-2066

[45] López-Julián PL, y Mandado-Collado JM. Extracciones químicas secuenciales de metales pesados: Aplicación en ciencias geológicas. Estudios Geológicos 2002;**58**:133-144

[46] Wang Z, Wang WY, Chen L, Yan C, Yan Y, Chi Q. Assessment of metal contamination in coastal sediments of the Maluan Bay (China) using geochemical indices and multivariate statistical approaches. Marine Pollution Bulletin. 2015;**99**:43-53

[47] Tessier A, Campbell PG, y Bisson M. Sequential extraction procedure for the speciation of particulate trace metals. Analytical Chemistry 1979;**51**:844-850

[48] Bloom NS, Preus E, Katon J, Hiltner M. Selective extractions to assess the biogeochemically relevant fractionation of inorganic mercury in sediment and soils. Analytica Chimica Acta. 2003;**479**:233

[49] Ure AM, Quevauviller PH, Muntau H, Griepink B. Speciation of heavy metal in soils and sediments. An account of the improvement and harmonisation of extraction techniques undertaken under the auspices of the BCR of the Commission of the European Communities. International Journal of Environmental Analytical Chemistry. 1993;**51**:135-151

[50] Mossop KF, Davdison CM. Comparation of original and modified BCR sequential extraction procedures for the fractionation of cooper, iron, lead, manganese and zinc in soils and sediments. Analytica Chimica Acta. 2003;**478**(1):111-118

[51] Meers E, Samson R, Tack FMG, Ruttens A, Vandegehuchte M, Vangronsveld J, Verloo MG. Phytoavailability assessment of heavy metals in soils by single extractions and accumulation by *Phaseolus vulgaris*. Environmental and Experimental Botany. 2007;**60**:385-396

[52] Milicevic T, Reli D, Skrivani S, Tesic Z, Popovic A. Assessment of major and trace element bioavailability in vineyard soil applying different single extraction procedures and pseudo-total digestión. Chemosphere. 2017;**171**:284-293

[53] Poggio L, Vrscaj B, Schulin R, Hepperle E, Ajmone-Marsan F. Metals pollution and human bioaccesibility of toposoils in Grugliasco (Italy). Environmental Pollution. 2009;**157**: 680-689

[54] Peijnenburg W, Jager T. Monitoring approaches to assess bioaccessibility and bioavailability of metals: Matrix issues. Ecotoxicology and Environmental Safety. 2003;**56**:63-77

[55] Ruby MV. Bioavailability of soil-borne chemicals: Abiotic assessment tools. Human and Ecological Risk Assessment. 2004;**10**(5):647-656

[56] Misenheimer J, Nelson C, Huertas E, Medina-Vera M, Prevatte A, Bradham K. Total and bioaccessible soil arsenic and lead levels and plant uptake in three urban community gardens in Puerto Rico. Geosciences. 2018;**4**(43):1-11

[57] Integral, Empresas Públicas de Medellín, EEPPM. Proyecto hidroeléctrico Rio Grande II. Estudio de factibilidad: Informe Final; 1982

[58] Morgan JJ. Manganese in natural waters and earth's crust: Its availability to organisms. In: Sigel A, Sigel H, editors. Metal Ions in Biological Systems. Vol. 37. New York: Marcel Dekker; 2000. pp. 1-34

[59] IUPAC. "Heavy metals"–a meaningless term? (IUPAC technical report). International union of pure and applied chemistry chemistry and human health division clinical chemistry section, commission on toxicology. Pure and Applied Chemistry. 2002;**74**(5):793-807

[60] Berner R. A new geochemical classification of sedimentary environments. Journal of Sedimentary Petrology. 1981;**51**(2):359-365

[61] Donahoe RJ, LIU C. Porewater geochemistry near the sediment-water interface of a zoned, freshwater wetland in the southeastern United States. Environmental Geology. 1998;**33**:143-153

[62] Pakhomova SV, Hall POJ, Kononets MY, Rozanov AG, Tengberg A, Vershinin AV. Fluxes of Iron and manganese across the sediment – Water interface under various redox conditions. Marine Geochemistry. 2007;**107**:319-331

[63] Urban NR, Gorham E, Underwood JK, Martin FB, Gordon OJ. Geochemical processes controlling concentrations of Al, Fe, and Mn in Nova Scotia lakes. Limnology and Oceanography. 1990;**35**(7):1516-1534

[64] Filius JDL. Adsorption of fulvic acid on goethite. Geochimica Et Cosmochimica Acta. 2000;**64**(1):51-60

[65] Weng LPK. Interactions of calcium and fulvic acid at the goethite-water interface. Geochimica et Cosmochimica Acta. 2005;**69**(2):325-339

[66] Sawyer DT. Oxygen Chemistry. New York: Oxford Press; 1991

[67] Salazar-Giraldo JP, Alfaro-De la Torre MC, Aguirre-Ramirez NJ, Briones-Gallardo R, Cedeño CJ, Peñuela-Mesa GA. Geochemical fractionation of manganese in the riogrande II reservoir, Antioquia, Colombia. Environment and Earth Science. 2013;**69**:197-208

[68] Rubio B, Nombela MA, Vilas F. Geochemistry of major and trace elements in sediments of the Ria de Vigo (NW Spain): Assessment of metal pollution. Marine Pollution Bulletin. 2000;**40**(11):968-980

[69] Buchman MF. NOAA Screening Quick Reference Tables, 2008 NOAA OR & R Report 08-1, Seattle WA, Office of Response and Restoration Division, National Oceanic and Atmospheric Administration, p. 34

[70] Hlavay J, Polyak K, Bodog I, Csok Z. Importance and determination of chemical species in solid simples. Micro Chemical Journal. 1995;**51**:53-63

[71] Salazar-Giraldo JP, Aguirre-Ramirez NJ, Peñuela-Mesa GA. Spatial and temporal behavior of soluble manganese at the sediment–water Interface in the Riogrande II reservoir in Colombia. Earth Interactions. 2015;**19d** Paper No. 11. ISSN. 1087-3562

[72] Rao CRM, Sahuquillo A, López-Sánchez JA. Review of the different methods applied in environmental geochemistry for single and sequential extraction of trace elements in soils and related materials. Water, Air, and Soil Pollution. 2008;**189**:291-333

Chapter 3

Extraction and Fractionation of Polysaccharides from a Selected Mushroom Species, *Ganoderma lucidum*: A Critical Review

Yew-Keong Choong, Kavithambigai Ellan,
Xiang-Dong Chen and Shaiful Azuar Mohamad

Additional information is available at the end of the chapter

http://dx.doi.org/10.5772/intechopen.78047

Abstract

Fractionation plays a big role in most of the sample processing especially in mushroom polysaccharides extraction. This intermediate step will determine further purification process which will lead to the type of polysaccharides that will be obtained. Four types of *Ganoderma lucidum* cultured medium used in the research papers were randomly chosen. They are spores, mycelia, fruiting body and fermentation broth. For water soluble polysaccharides, hot water extraction is typically applied. The following ethanol precipitation could be appropriate used to sediment the component with OH-group including polysaccharide. The next step of fractionation consist of anion exchange chromatography or gel filtration enhance the purity of polysaccharides. Using these extraction and fractionation techniques, high quality polysaccharides could be successfully obtained from the mushroom that are useful for further studies. This review examined the various extraction and fractionation techniques used in the study of polysaccharides from *G. lucidum*.

Keywords: extraction, fractionation, polysaccharide, *Ganoderma lucidum*, anion exchange chromatography

1. Mushroom polysaccharides

Mushroom polysaccharides have been studied intensively since the therapeutic effects of hot water extract from several fungi were evaluated on Swiss albino mice [1]. The fungi, *Pleurotus ostreatus, Pholiota nameko, Pleurotus spodoleucus, Tricholoma matsutake, Flammulina velutipes* and *Lentinula edodes* have been shown to potentially inhibit the growth of transplanted Sarcome 180 tumor cells [2]. These significant results have prompted many scientists to purify and isolate the therapeutic compounds. One of the effective constituents was found to be β-D-glucan. It is one of the main constituents of medicinal mushroom, the main chain consisting of β-(1→3) linkages with some β-(1→6) branches as well as chitin, mannons, galactans, and xylans. **Figure 1** shows the molecular structure of β-D-glucan.

The long chain of main β-glucan molecules is capable of having branching side-chains with other monosaccharides in the position of C_1 and C_3 or C_1 and C_6 through condensation. Likewise, other types of molecules such as proteins are found in polysaccharide-K (**Figure 2**) [3].

Basically D-glucose units are the common forms of β-glucans and mostly they are binding with β-1,3 links. The β-glucans of mushroom, yeast or other fungi contains 1–6 side branches (**Figure 3**), while cereal β-glucans contain β-1,4 and β-1,3 backbone bonds.

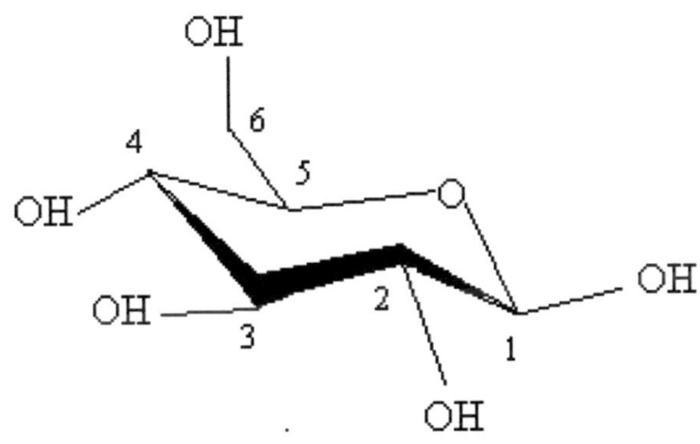

Figure 1. Six-sided D-glucose rings connected linearly at varying carbon positions. When the —OH group at C_1 direct 45° toward C_3 at another molecule glucose to have glycosidic bond β(1→3) and the continuing linkage forming the larger polymer.

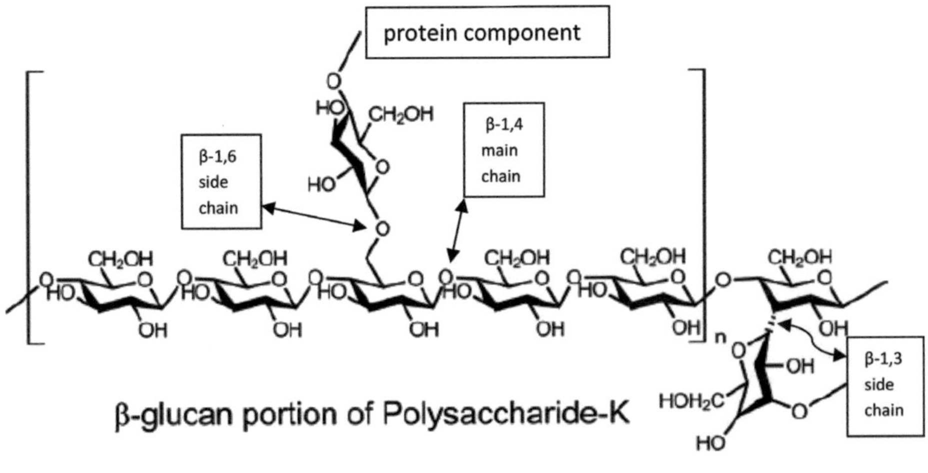

Figure 2. Side-chains of β-glucan with protein in polysaccharide-K. Polysaccharide-K (Krestin, PSK) is a protein-bound polysaccharide purified from the fruiting body of *Coriolus versicolor*. PSK predominantly consists of a beta-glucan β-1,4 main chain with β-1,3 and β-1,6 side chains. The molecular weight of PSK is 100,000 Da, its approximately 25% tightly bound protein is reported at the β-1,6 side chain.

http://dx.doi.org/10.5772/intechopen.78047

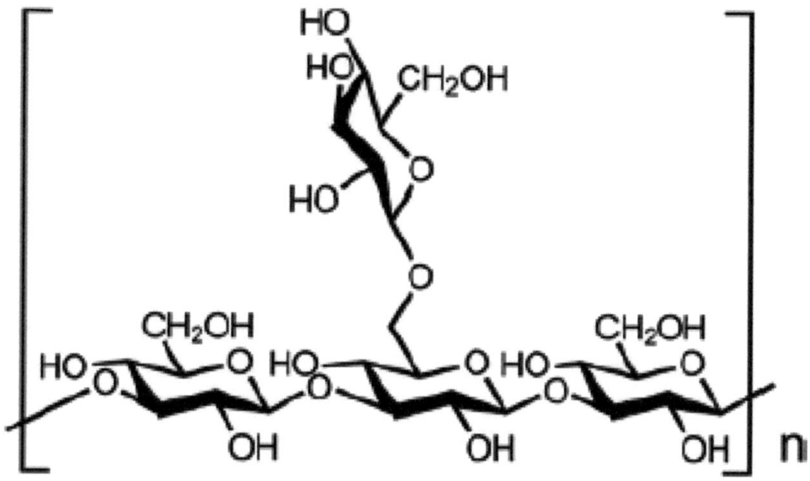

Figure 3. Structure of backbone β-1,3-glucan with β-1,6 branching. n is the alphabet numbers.

In short, there are many possible branches of different types of monosaccharides with glycosidic linkages. This scenario is closely associated with molecular weight of the polysaccharide.

Presently the effects of mushroom polysaccharides have been studied intensively which consist of antioxidant [4], anti-inflammation [5], anti-hyperglycemia [6], anti-tumor [7] and immunostimulation [8]. Nevertheless the process of discovery of new polysaccharides is still on-going. Moreover, research revealed that the chemical characteristics of the constituents of the mushroom polysaccharides, such as molecular weight, molecular structure, solubility in water, the location, length and frequency of the side-chains have a correlative relationship with the function of polysaccharides [9]. The issue of types of isolated polysaccharides have received considerable critical attention on the step of the process of extraction especially fractionation. Fractionation has a pivotal role in the intermediate stage between extraction and isolation. Polysaccharides fractionation is essential for a wide range of technologies for separation process including certain quantity of a crude polysaccharides. The crude will be continuously separated into smaller quantities with diverse compositions.

2. Polysaccharides of selected species, *Ganoderma lucidum*

Ganoderma lucidum is one of the popular medicinal mushrooms. It belongs to the basidiomycetous and assigned to the family *Polyporaceae*. It has been used in Chinese traditional medicine for more than 4000 years [10]. This fungus with long medicinal history has been studied intensively. Typically such studies included its genetic profiles, chemical composition, multiple physiological testing, toxicology and clinical trials. The bioactive compounds especially the polysaccharides of this mushroom generate much attention due to its host dependent action [11] and involving immune stimulation system [12].

Polysaccharides of *G. lucidum* have been studied and are found typically to contain the pure polysaccharides, but some are also binding with proteins or peptides. It was previously reported that the molecular weight of these polysaccharides varied widely, e.g., a polysaccharide peptide with a molecular weight of 584,900 Da was isolated from a hot water extract of wood-cultured *G. lucidum* by Cao *et al.* [13]. It was found to contain more than 40% β-D-glucan. Another report by You *et al.* showed a type of polysaccharide peptide extracted from the same species with an average molecular weight of 513,000 Da. It contains 16 kinds of amino acids [14]. Most of the polysaccharides extracted and bonded with protein had high molecular weight and played a specific function related to immune stimulation system. It was reported by Giavasis [15] and Ho *et al.* [16] that bioactive polysaccharides which are higher in molecular weight have been shown to exhibit significant antitumor properties, immunomodulatory activity, antioxidant activity, and neuroprotection. A series of fractionation has been conducted with the purpose of studying this little known fungus using advanced technology. A pure *G. lucidum* polysaccharide labeled as GLP-1-1 was isolated from a culture broth with molecular weight of 22,014 Da [17]. Its monosaccharide content contained glucose, mannose, and galactose with molar percentages of 92.33, 7.55, and 0.22%, respectively. A water-soluble polysaccharide [18] extracted from the spores of *G. lucidum,* subsequently went through water extraction and sequential alcohol precipitating method and purified by anion-exchange and gel filtration chromatography. It was eluted as a single and symmetrical sharp peak corresponding to an average molecular weight of 8000 Da as determined by high performance gel permeation chromatography. Sugar compositional analysis showed that it is only composed of D-glucose. Wang *et al.* [19] used a native *G. lucidum* polysaccharide in a linear, water-insoluble β-D-(1→3)-glucan to prepare sulfated polysaccharides. The molecular weight of *G. lucidum* polysaccharide was 133,000 Da. In addition, another water-soluble neutral polysaccharide was isolated from the fruiting bodies of *G. lucidum* by DEAE Sepharose Fast Flow and Sephacryl S-500 High Resolution Chromatography [20]. GC analysis showed that this polysaccharide was mainly composed of glucose and galactose in the molar ratio of 34:1 and its average molecular weight was approximately 2,500,000 Da. Contradictorily, there are two low-molecular weight of glucan purified from a crude *G. lucidum* polysaccharide preparation [21]. Their physiochemical properties marked a glucan with 5200 Da and another glucan with molecular weight 15,400 Da. The later was composed of glucose, galactose and mannose in a ratio of 29:1.8:1.0. This wide range of molecular weight of *G. lucidum* polysaccharides apparently clearly indicated that they are still abundant polysaccharides of *G. lucidum* yet to be discovered.

The molecular structure of *G. lucidum* polysaccharides which is associated closely with glycosidase linkage is another factor vitally influencing the bioactivities. The types of monosaccharide used to be the backbone play an important role as skeleton to support the other types monosaccharide as branches. Actually, it is a complex polymer upon which the different types of monosaccharides were involved plus the different position of carbon binding. For example, two peptidoglycans named Ganoderan B and C isolated from the fruiting body of *G. lucidum,* were shown to exhibit hypoglycemic efficacy with molecular weight 7400 and 5800 Da [22]. Chemical and physicochemical studies demonstrated that the backbone of ganoderan B contain D-glucopyranosyl β-1→3 and side chains with β-1→6 linkages. The backbone of Ganoderan C contains glucopyranosyl β-1→3 and side chains with β-1→6 linkages and a D-galactopyranosyl β-1→6 linkage.

3. Review of available *G. lucidum* polysaccharides extraction and fractionation methodology

Mushroom polysaccharides are present as structural components of fungal cell wall which is composed of two major types of polysaccharides. They are divided into a rigid fibrillar of cellulose and a matrix-like glycoprotein, α-glucan or β-glucan [23]. Selection of mushroom polysaccharides extraction method normally relies on the cell wall structure. A reliable procedure for successful extraction of polysaccharides from either cultivar mycelia or fruiting body has been developed [24]. The extraction method commonly involves 80% ethanol for elimination of low molecular substances from mushroom material, followed by 3–5 successive repeating extractions with water (100°C, 2–4 h). The alternative choice is 5% sodium hydroxide (80°C, 6 h) or 2% ammonium oxalate (100°C, 6 h). In fact, the hot water extraction yields water-soluble polysaccharides, while the extraction with alkali solution is the best method to produce water-insoluble polysaccharides. Although various types of extraction method can be applied depending on the structure and water-solubility of polysaccharides, however, it is important to rupture the hard cell wall from the outer layer to the inner layer with weak-to-strong extraction conditions (pH and temperature). A combination of techniques should be chosen for further extraction of polysaccharides, normally ethanol precipitation will be the first choice as it excludes the impurities from the extracted polysaccharides. Other techniques included fractionation precipitation [20], acidic precipitation with acetic acid [21], ion-exchange chromatography [22], gel filtration [23] and affinity chromatography [24].

Comparing to the ancient era, definitely today's new technology is enhancing the yields of the extraction. A lot of new polysaccharides of *G. lucidum* have been discovered and reported to possess certain level of therapeutic efficacy [25, 26]. However, the method of polysaccharides extractions of this mushroom including the fractionation performed by the researchers are not standardized. Numerous methods resulted in the extraction of incorrect polysaccharides especially when using the hot water extraction method. The objective of this review paper is to discuss several polysaccharides extraction methods and verify these methods which include fractionation on the selected mushroom, *G. lucidum*.

3.1. Extraction and fractionation of a novel water-soluble β-D-glucan from the spores of *G. lucidum* using ethanol precipitation

Analysis was based on the conceptual framework proposed in five papers authored by Bao *et al.* from 2000 to 2002 [27] (please refer to **Figure 4**). The subject model was the spore of *G. lucidum* which was sporoderm-broken. With reference to the paper by Bao *et al.* [28], a novel water-soluble β-D-glucan was discovered after 10 years of study. Most of the steps of the extraction and fractionation were similar except that several procedures were modified in order to obtain better results. This illustrated that older methods could still work well and these methods could give rise to more effective methods that could lead to new discoveries.

Normally 95% ethanol is added at the beginning of extraction to remove the lipids especially after the sporoderm of the spores were broken. The exposure to the environment actually escalated the oxidation of the content of spores. Two types of spores breaking mechanisms are applicable nowadays. The traditional way, i.e., via mechanical vibration grinding resulted in the bioactive ingredients and unsaturated oil being directly exposed to the air, while the Supersonic Airflow Pulverizer would only break the outside hard chitinous layer and keep the inner covering layer complete [29]. The methodology in this study did not mention the types of spores breaking mechanism and storage period of sporoderm-broken spores before starting the extraction.

The defatted step, conducted twice, 5 days duration in each, would remove maximum lipid contents. After centrifugation, the residues, with most of the ethanol content being removed, have to be air-dried to ensure complete ethanol free. The hot water extraction is particularly useful to

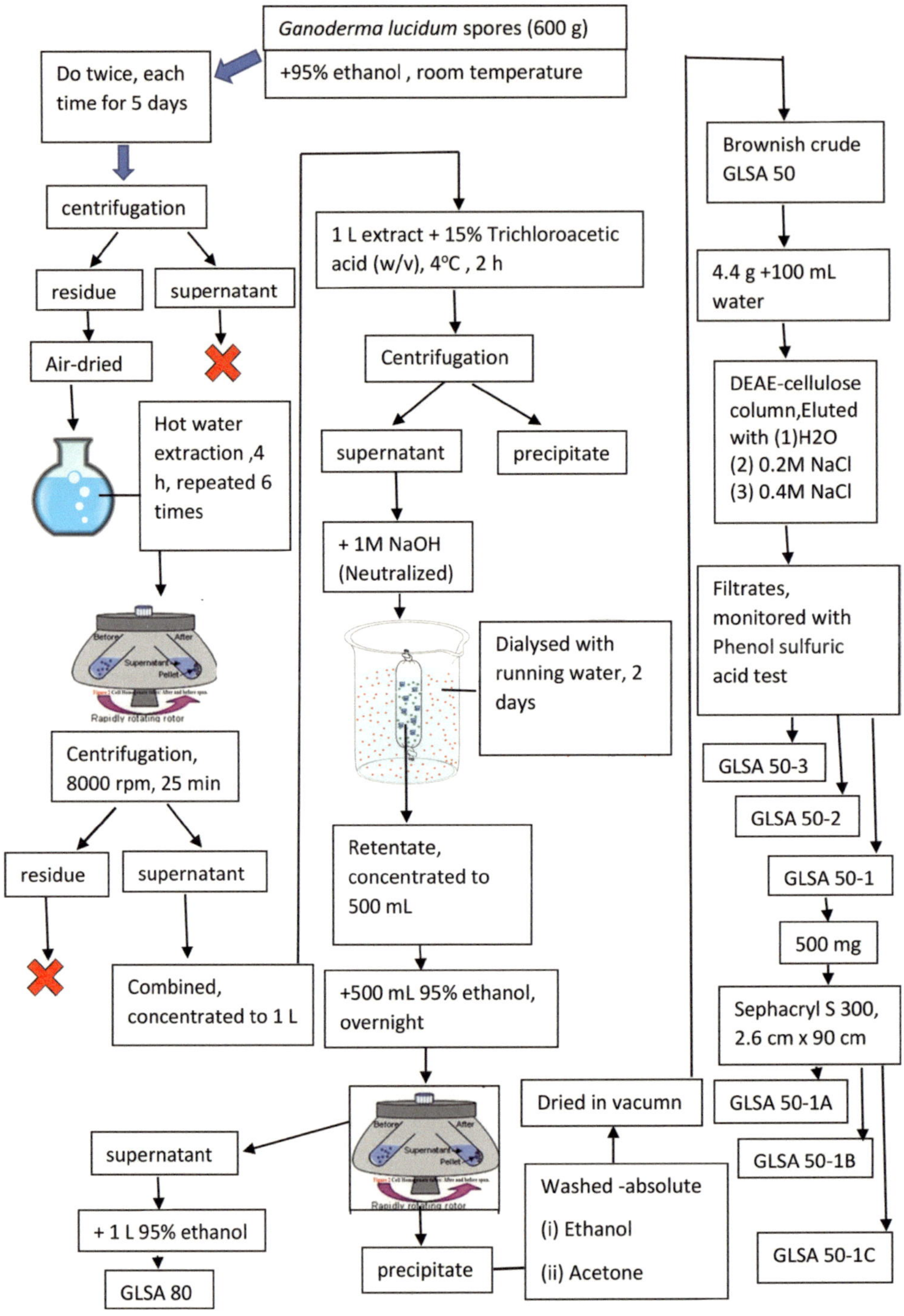

Figure 4. Flowchart of extraction, fractionation and isolation methodology of polysaccharides from *Ganoderma lucidum* spores.

extract water-soluble polysaccharides in the mushroom. Six replicates of extraction with boiling water with interval of 4 h each would allow the changing of new batch of hot water. This suggested that the bigger spin probe provided at the center of the container with appropriate speed would reduce the bubbles and increase the volume of spores with water. The extraction of substances with OH group under power of polarity of water and temperature could ensure the effective extraction of reishi spores polysaccharides. Therefore the accumulated 60 L of hot water filtrate needs to be centrifuged until 1 L of concentrated filtrate is obtained for further processing.

For this study, 15% of trichloroacetic acid (TCA) was used to procure protein precipitation reaction [30]. In fact, most of the protein components have been denatured during boiling. Continuously added TCA eliminated components that interrupted the polysaccharide purification process. The next step was dialysis of the supernatant with 1 M of sodium hydroxide aimed to neutralize the TCA. The dialysis was completed after 2 days under running water. However, Bao *et al.* [28] reported using 3 days of running water followed by 1 day of distilled water to remove components with molecular weight of 3000 Da. Later the concentrated retentate was stirred slowly and 95% ethanol was added at low temperature to promote polysaccharides precipitation. After centrifugation the precipitate was washed with two solvents before being dried in vacuum. Compared to previous study [28], the precipitate obtained was dissolved in water at room temperature, 0.5 L of ethanol was added and another round of centrifugation was conducted. The resulting brownish crude product was the key component used for further fractionation. The packed column with DEAE-cellulose (diethylaminoethyl) was used in anion-exchange chromatography. It achieves separation by using positively charged ion exchange matrix with an affinity for molecules having net negative surface charges [31]. The elution by using water and various affinity salts would be adequate to dissolve the polysaccharide components which are retained in the matrix. The fractions collected in tube were colorless, and the best determination method was the phenol-sulfuric acid method [32], since the yellowish end product could be used for the calibration graph based on absorbance at 490 nm versus number of fractions. The peak determined from the graph could be used to determine combination of fractions for further purification. Thus sephacyl-S-200HR [33] is used as size exclusion ordinal isolating each different polysaccharides based on their molecular weight. In this case, the compound labeled as GLSA50-1B was chosen for the characterization as a novel β-D-glucan.

The key aspect of this experiment was isolation of the pure polysaccharides and not total carbohydrates. Hence the definition of polysaccharide needs to be fully understood. There was confusion on the nature of polysaccharides when the experiment was terminated at the time when only accumulated filtrates were obtained after a few rounds of hot water extractions. The claim that polysaccharide could not be substantiated, when steps included protein-precipitation, dialysis, ion-exchange chromatography and size exclusion, were ignored. The center concept of polysaccharide is that it consists of the skeleton of the sugar backbone structure, the glycosidase linkage of branches, and specific rotation to indicate the configuration. The result of this study provided further support of new and significant new polysaccharides from mushroom spores.

3.2. Extraction and fractionation of *G. lucidum* polysaccharide from the mycelia of *G. lucidum* using agitation and filtration

The topic has mentioned "polysaccharides" which implied many types of polysaccharides. and this was frequently attributed to hot water extraction of mushroom resources [34] (please refer to **Figure 5**). The methodology described seems short and insufficient to determine the major characteristics of polysaccharides. To date, studies investigating *G. lucidum* polysaccharides have produced equivocal results that contributed to misunderstanding of polysaccharides. Firstly, when the extraction steps were carried out using hot water extraction and terminated after precipitation with 95% ethanol, most of the claimed polysaccharides (labeled as GLP) comprised total carbohydrates. In addition, majority of the polar compounds with plenty of –OH groups could possibly have been evacuated from the matrix. The experiment was unable to encompass the entire full scope of detailed molecular level of specific polysaccharides. Secondly the composition of GLP included alduronic acid (19.27%) and protein (5.39%). This scenario directly indicated the GLP content was not purely polysaccharides. The objectives of this study focused on efficacy of GLP treatment in plasma insulin concentration and gut microbiota composition in mice. There is notable paucity of empirical research focusing especially on polysaccharide.

The extraction started with pulverized *G. lucidum* mycelia with water. The process was repeated twice and the mixture was agitated for 4 h at 70–80°C. Comparing the common hot water extraction which provided 100°C of boiling water, it can be seen that the reduced temperature used was to avoid the formation of bubbles. The agitation was accomplished by continued shaking for 4 h. These mechanical activities are adjustable to ensure the extraction would be maximum. However, there is no mention on the speed of shaking in the experiment and the condition of the mycelia.

A variety of perspectives were expressed at the end of study. The striking result to emerge from the data was the positive and significance of GLP treatment. However, this finding cannot be extrapolated to determine the effectiveness of polysaccharides due to the impurity of GLP.

3.3. Extraction and fractionation of *G. lucidum* polysaccharides F31 from the fruiting body of *G. lucidum* using absolute ethanol at 4°C

Comparing with the common hot water extraction, this study applied 80°C instead of 100°C water [35] (please refer to **Figure 6**). Recent trends using temperature lower than 100°C have led to the preservation of protein content and elimination of bubble forming. One of the greatest challenges is the extraction of the polysaccharides with hot water, concurrently preventing the denaturation of the protein content. Typically, when the temperature is above 41°C, the protein will probably be melted but most of the peptides maintain their primary structures. Questions have been raised about the degree of denaturation of protein at temperature below 80°C compared to 100°C. Eventually the secondary and tertiary structures of protein will be destroyed under higher temperature. Heat could be a contributing factor for the production of end products during the process of water extraction. In this case, protein analysis had been

http://dx.doi.org/10.5772/intechopen.78047

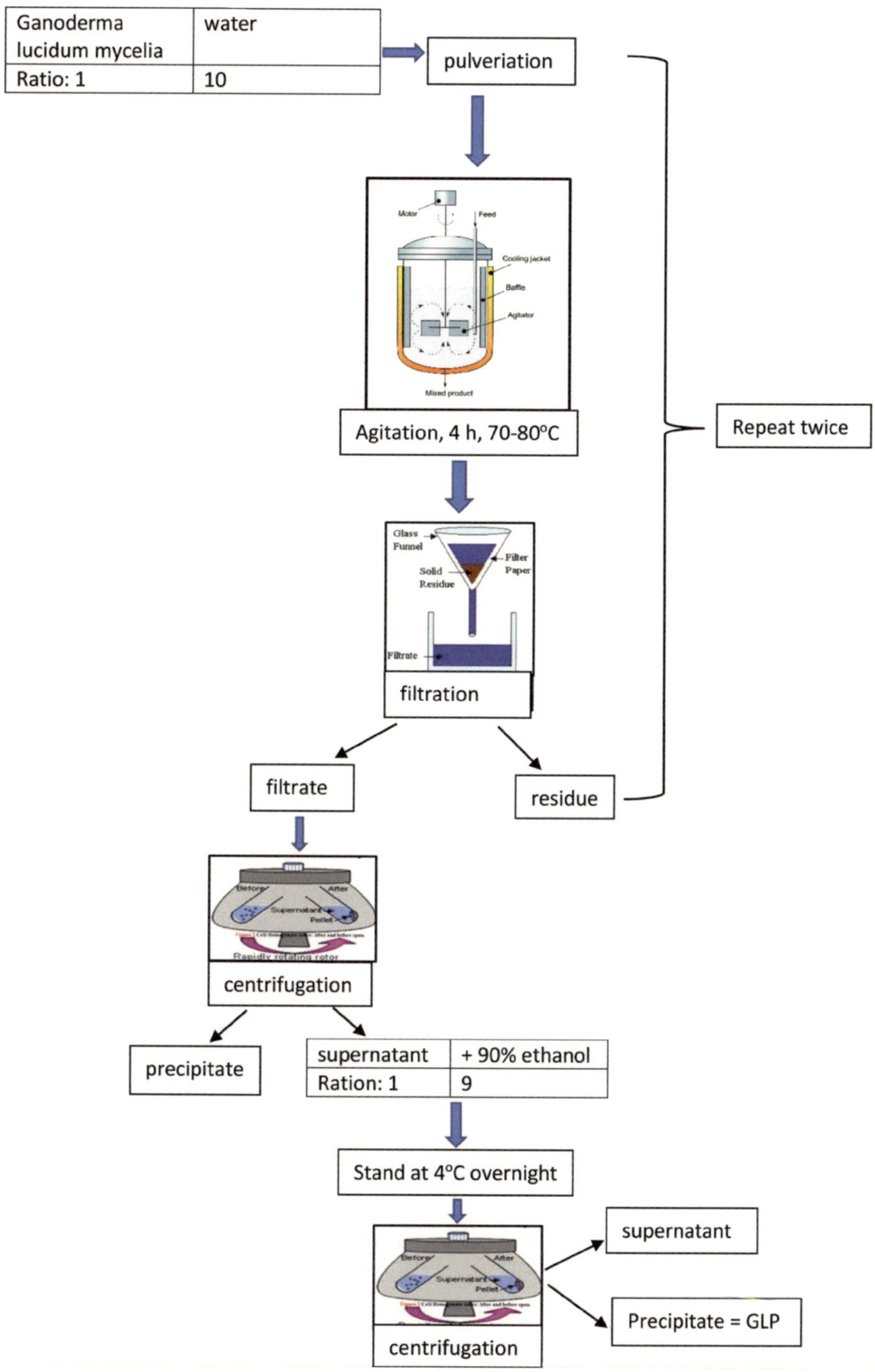

Figure 5. Flowchart of extraction, fractionation methodology of polysaccharides from *Ganoderma lucidum* mycelia.

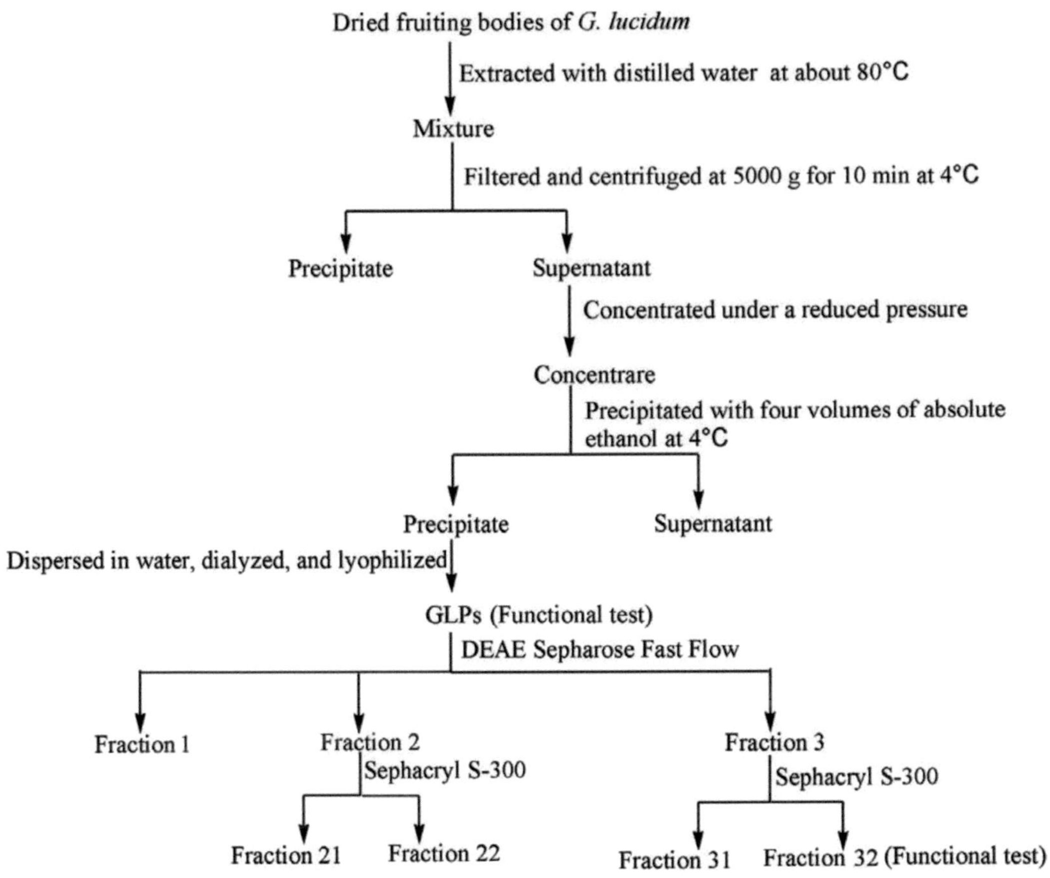

Figure 6. Flowchart of extraction, fractionation and isolation methodology of polysaccharides from *Ganoderma lucidum* fruiting body. The gel DEAE Sepharose Fast Flow and Sephacryl S-300 were the main tools to obtain the pure polysaccharides.

carried on during the process of extraction. Despite the fact that the main objective of the study was to accentuate the polysaccharides extraction, certain conditions were required to determine of protein or peptide content.

After the precipitation of GLPs by DEAE-sepharose Fast Flow, fractionation was conducted using affinity chromatography. The fractionation has been applied intensively with the purpose of obtaining single pure compound. The subsequent usage of Sephacryl-S-300 resin [36] as high resolution size exclusion chromatography allows rapid and reproducible purification of polysaccharides, proteins and other macromolecules. In this study, this resin purified the polysaccharides with fractionation range of globular polysaccharides size 1×10^4 to 1.5×10^6 Da.

The results showed that P31 has been successfully fractionated and characterized with subsequent determination of molecular weight. The different techniques of analytical results indicated that P31 is a pure compound. Thus, this study showed an example of obtaining pure polysaccharides for bioassay guided test.

3.4. Extraction and fractionation of high molecular weight bioactive-β-glucan from the fruiting body of *G. lucidum* using centrifugation and high performance anion exchange chromatography

Previous studies have reported that the fractionation for polysaccharides from mushroom sample normally started with hot water extraction [37] (please refer to **Figure 7**). This initial step, albeit old, ensures the polysaccharides are the main compounds obtained at the end of the extraction. The function of polysaccharide is largely based upon empirical studies that demonstrated a strong and consistent association with immune system of host.

Similar procedure of hot water extraction was used on dried dark brownish powder form of fruiting body and freeze-dried white powder (labeled GLP20 in this paper), compared to procedure for mycelia and spores. The minor differences of the temperature, duration of hot water extraction, speed and time of centrifugation and volume of ethanol could be due to the larger sample of volume being used.

The fractionation of GLP20 started when 2 M trifluoroacetic acid (TFA) was added at 110°C for 3 h. This colorless strong acid has a higher acid ionization constant compared to acetic acid and trichloroacetic acid, due to the fluorine atom exerting the inductive effect that strategically stabilizes the CO^{2-} charge of the carboxylic anion more than the other corresponding acids [38, 39]. Therefore the glycosidic linkage of GLP20 was broken down and hydrolyzed to monosaccharides composition. One possible implication of this is the separation of the sugar units of the polysaccharide as its large molecular structure, and later the determination of the monosaccharides which can be done with analytical column. However, another issue has to be considered. GLP20 was the product of the precipitation by ethanol. In this study, open column for further purification of GLP20 was not used, despite the fact that references were cited in the section of results on the purification by columns such as DEAE cellulose chromatography and Sephadex series size exclusion chromatography. The authors commented that this process was tedious and the yield of purified polysaccharides was low. The present studies yielded 0.37% (w/w) GLP20 on the basis of the dry weight of fruiting bodies. The mean 5.55 g of GLP20 was obtained by using 1500 g fruiting body of *G. lucidum*. If each batch of analysis needed 2 mg of GLP20, a total of 2775 times of analysis can be conducted.

The present study was over-reliance on self-initiated methodology. The advantages of using a series of chromatography for isolation at the final stage of fractionation are for the purpose of maximizing the purification. The use of 20% ethanol precipitation was supposed to precipitate all substrates with —OH group. Even at the higher % of ethanol applied, there are increasing precipitation of substrates including polysaccharides.

The degree of purification of GLP20 was uncertainly proven. Thus the content of monosaccharides reported later did not appear to support the assumption of single polysaccharide obtained from this extraction, since the process of hydrolysis was not specific to determine the bonding of each sugar component. Generally, hydrolysis or saccharification is a step in the degradation of a polysaccharide substance. The cleavage of H—O bond reaction of cation and anion or both with water molecule takes place due to pH conditions [40]. It is possible that these reports were influenced by lack of information on glycosidic linkages. The final analysis step included total protein test.

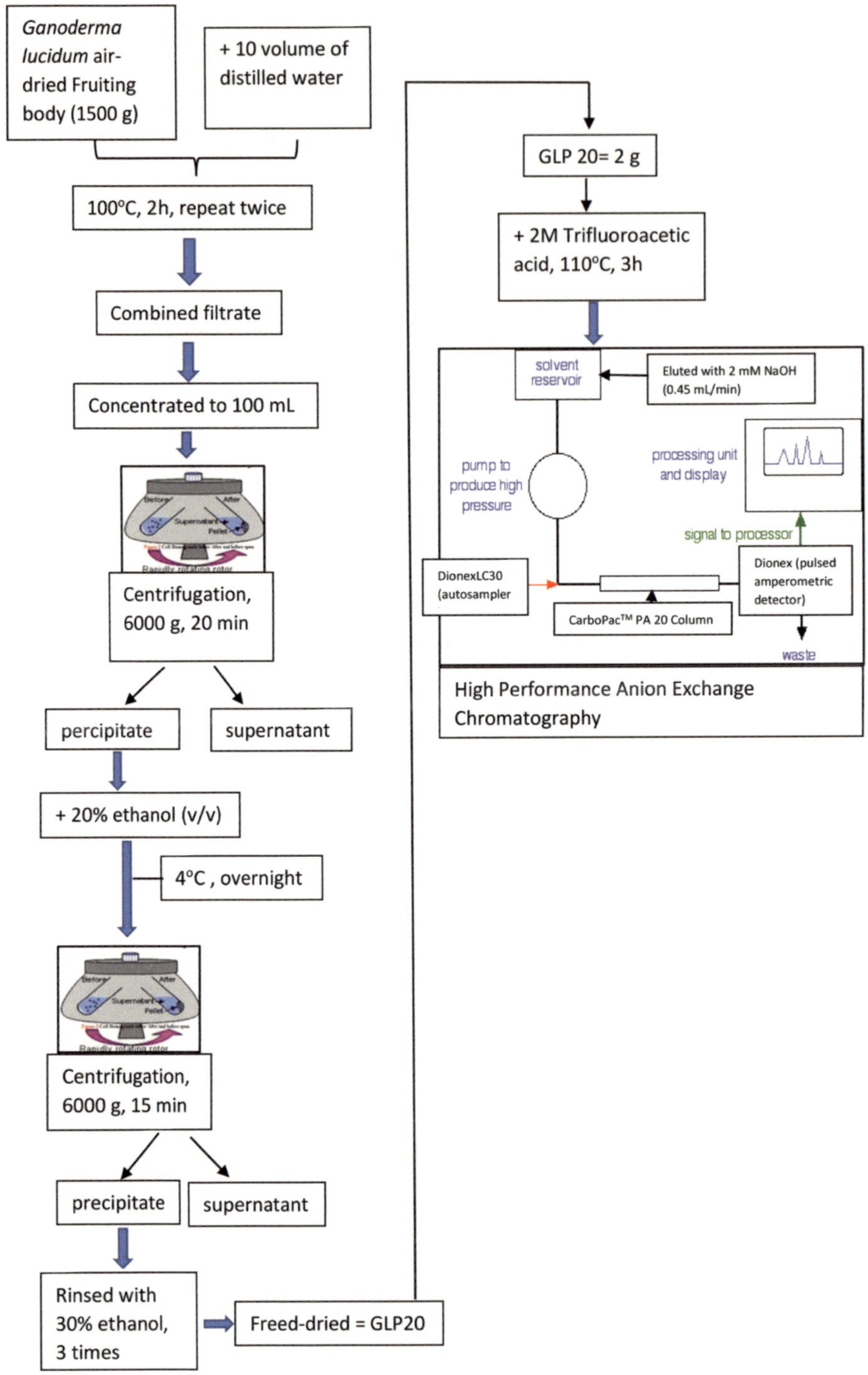

Figure 7. Flowchart of extraction, fractionation and isolation methodology of polysaccharides from *Ganoderma lucidum* fruiting body.

3.5. Extraction and fractionation of mushroom polysaccharides from the mycelia and fruiting body (dried of *G. lucidum* and *Poria cocos* using soxhlet extractor and anthrone sulfate)

Two types of mushroom have been used [41] (please refer to **Figure 8**). They were *G. lucidum* and *Poria cocos*. The methodology used was similarly applied on both mushrooms. Sample mushrooms were in the form of mycelia and fruiting body, respectively. In other words, the method aimed to produce total polysaccharide as mentioned in the article and there was no specific technique to produce targeted polysaccharide from either mycelia or fruiting body. There was no indication whether the samples were in fresh or dried condition, and it was assumed that 2 g of dried powder sample plus 90 mL of distilled water processed in soxhlet extractor. The final concentrated extract after distillation for few hours was diluted with distilled water. This step is important because the final reading of absorbance at 625 nm is limited at 3. When 10% volume of diluted soxhlet extract was added with ethanol, it was possible that the micro-molecules that existed during the soxhlet heating were eliminated. Hence the next step of chilling at 4°C provided ample time for the suspension of the unwanted debris. The speed and duration of centrifugation were commonly applied to most of precipitate. After separation, the precipitate still needs to be diluted for determination of carbohydrate by the anthrone method. The principle of anthrone method is that dehydrated carbohydrates in various form included monosaccharides, disaccharides, starch, gums, glycosides, dextrins and definite polysaccharides during the process will condense with anthrone to form a green color complex which can be measured colorimetrically at 625 nm.

This soxhlet extraction method is one of the practical ways to exclude types of carbohydrate from the mushroom. Therefore, the polysaccharides mentioned in the chapter would refer to various forms of carbohydrates which included polysaccharides as one of them. The result and conclusion has positively quoted the significance of said production of polysaccharides potentially acting as prebiotics by manipulating gut microbiota composition. It was decided that the best method to adopt for this investigation was to provide a platform for beneficial bacteria associated with polysaccharides treatment, making these polysaccharides candidate prebiotics. However, there was a lack of result specifically on types of polysaccharides from both basidiomycetes.

3.6. Extraction and fractionation of three kinds of *G. lucidum* polysaccharides from the fermentation broth, mycelia, fruiting body using open column with resin DEAE-Sephadex A-25 and Sepharose 6B

Three different life stages of *G. lucidum* were involved in this study [42] (please refer to **Figure 9**). Since it takes around 3 months for cultivation of *G. lucidum* fruiting body, the mycelia-based and broth-based products have assumed extraordinary importance because of the requirement for heightened quality-control and year-round production [43]. The processes and the different types of growth parameters including temperature and pH concerned in submerged mycelial medium, are simple to be standardized under controlled circumstances. In addition, the purification and other succeeding processing of active components such as polysaccharides excreted into the culture medium normally involve relatively simple procedures. The cultivation

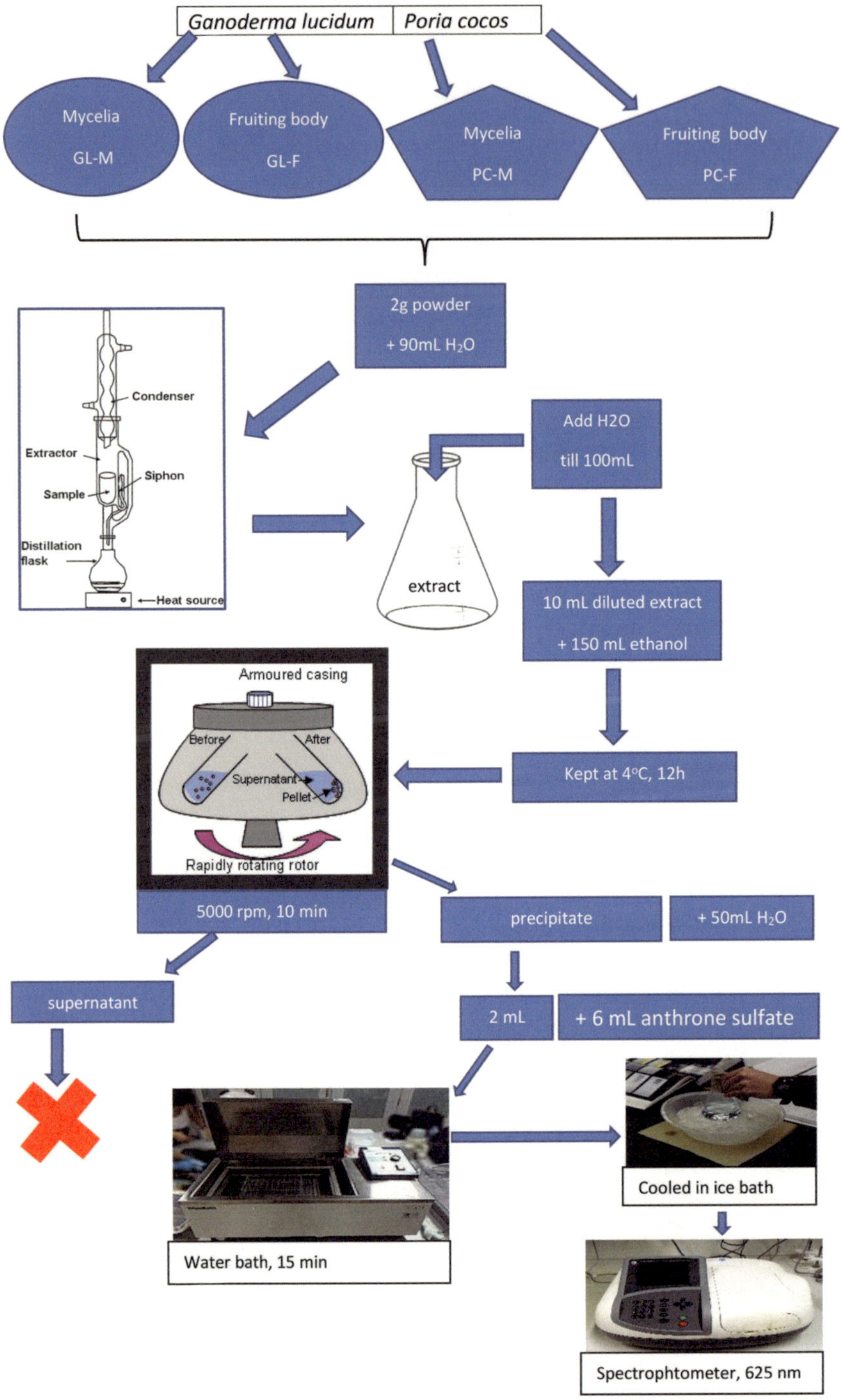

Figure 8. Flowchart of extraction, fractionation methodology of polysaccharides from *Ganoderma lucidum* and *Poria cocos*, fruiting body and mycelia.

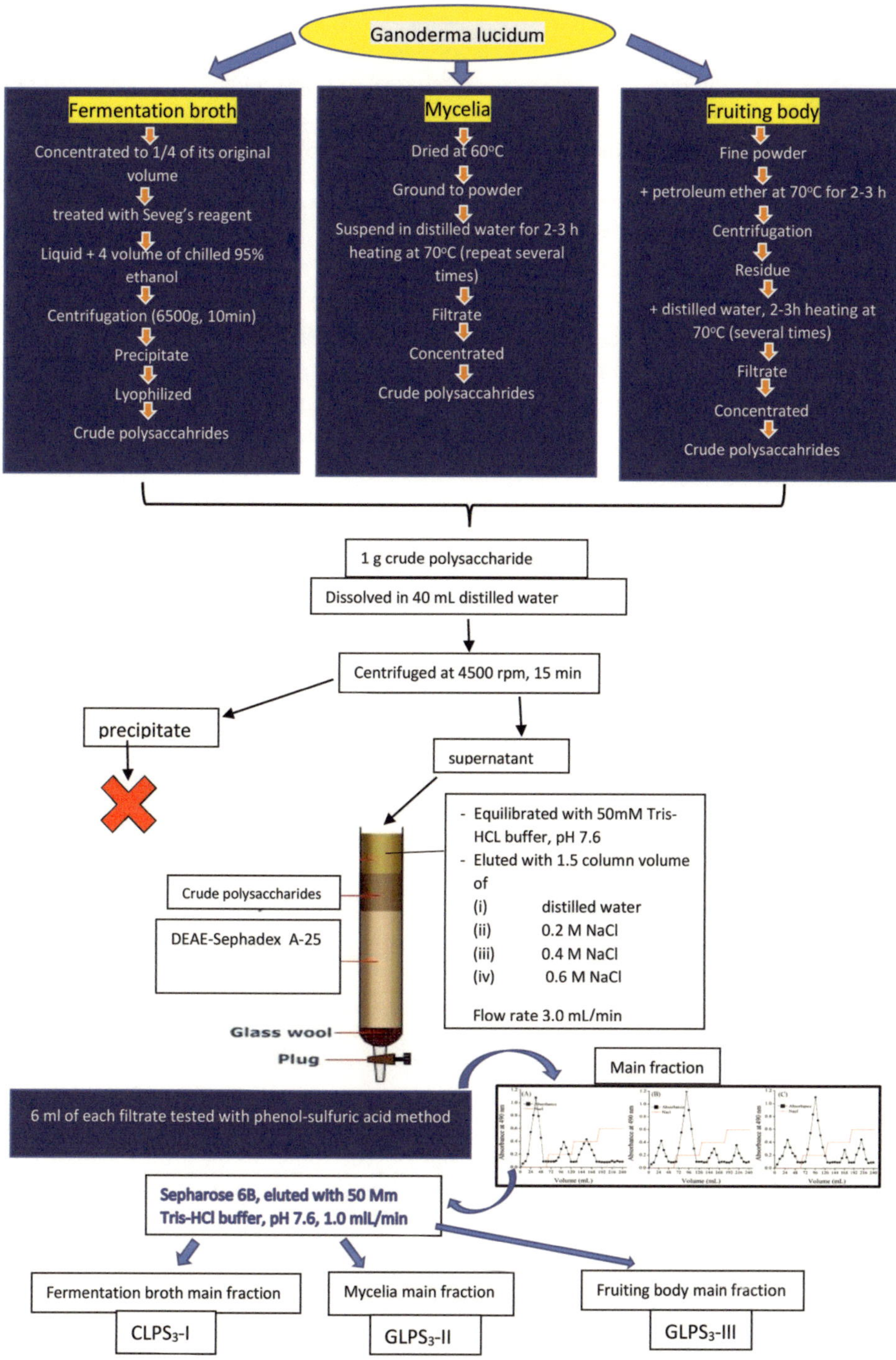

Figure 9. Flowchart of extraction, fractionation and isolation methodologies of polysaccharides from *Ganoderma lucidum* fermentation broth, mycelia and fruiting body.

environment and the ingredient of medium have also been reported to strongly impact the growth of mycelia and the production of exopolysaccharides [44]. A report by Yang and Liau [45] noted that production of *G. lucidum* polysaccharides by fermenter-grown mycelia cultivar was optimum at pH of 4–4.5 and temperature 30–35°C. Acceleration of mycelial growth and bioactive components could be produced by addition of supplements such as fatty acids.

Three types of dissimilar pre-treat ways were applied on the different life stages of mushroom before the hot water extraction. An investigation of the non-volatile composition of *G. lucidum* [46] reported that the mushroom contains 59% crude fiber, 26–28% carbohydrate, 7–8% crude protein, 3–5% crude fat and 1.8% ash. Nonetheless, a wide variety of bioactive molecules are also found in *G. lucidum* which include terpenoids, phenols, nucleotides and their derivatives, steroids, glycoproteins and polysaccharides. *G. lucidum* proteins contain most of the essential amino acids and are especially affluent in leucine and lysine. The fermentation broth contained media, mycelia and spores. After concentration, the volume of liquid was reduced. Addition of Sevag's reagent for the purpose of deproteinization could eliminate the protein ingredient in the mixture. Xia Li *et al.* [47] proposed an explanatory theory of the advantage of using Sevag's reagent. With regards to deproteinization rate and polysaccharide residual rate, the process of optimization, was based on the following parameters: deproteinization frequency (example: 4 times), time of oscillating (example: 11 min), the volume ratio of sample to Sevag's reagent (2:1), and the volume ratio of chloroform to n-butanol (5:1), The resulting deproteinization rate was 64.2%, the polysaccharide loss rate was 34.3%, and the purity of polysaccharide was raised by about 2.8 times. Compared to other stage of cultivation, mycelia of *G. lucidum* were treated simply and without any step of deproteination and removal of fatty substances. One of the reasons for these could be that at the initial stage of mycelial mat forming, little protein was available. Excluding the step of deproteination allowed the hot water extraction at 70°C to be conducted directly and shorten the time for processing polysaccharides. Similarly, deproteination for fruiting body powder is also not required. Instead, petroleum ether is used to remove fatty and fat-soluble substances. The high proportion of polyunsaturated fatty acids and low total fat content relative to the total fatty acids of *G. lucidum* are considered significant contributors to the heath value of *G. lucidum*. [48]. The residues accumulated after the defatted step would allow the hot water extraction to proceed.

In this case, the main concern at the end of experiment was isolation of polysaccharides. As in other studies, from the accumulated filtrates of several replicates of hot water extraction, the precipitate treated by addition of ethanol and freeze-dried would contain crude polysaccharides, indicating that several types of polysaccharides are found in the crude. The subsequent step in this study was the application of DEAE-Sephadex A-25 column as anion exchange chromatography intended for the further separation of each type of crude polysaccharides. According to the data file of ion exchange [49], Sephadex ion exchangers are derived from either Sephadex G-50 or Sephadex G-25. The G-50 matrix is less highly cross-linked than the G-25. Ion-exchangers based on Sephadex-50 have less rigidity and thus swell more than those based on G-25, which are more rigid. These properties mean that A-50 and C-50 types are better suited to larger biomolecules, such as polysaccharides or proteins in the molecular weight range of 30,000–100,000 MW, whereas A-25 and C-25 type ion exchangers are a better choice for small molecules up to about 30,000 MW.

In this stage of fractionation, Sephadex A-25 was used instead of A-50 and followed by another stage of fractionation with Sepharose 6B. Hence, most of the molecules more than 30,000 MW could pass through. In order to completely isolate the polysaccharides, the main fraction of each stage of *G. lucidum* was successfully eluted with 50 mM Tris-HCl buffer.

The excellent results presented in this study illustrated the effective methodology for purification of polysaccharides from *G. lucidum* using three different life stages. The dissimilar way of pre-treating the sample in the beginning has potentially separated all the polysaccharides. The results clearly showed each main isolated polysaccharide contained different total sugar content, monosaccharides composition, molecular weight and the bioactivity. This study had successfully isolated the main polysaccharides from fermentation broth, mycelia and fruiting body of *G. lucidum*.

4. Fractionation and its application in the extraction of polysaccharides from the mushroom species, *G. lucidum*

The discovery of new polysaccharides from *G. lucidum* has long been a research of great interest in a wide range of field. Bao *et al.* [50] reported six different functionalized derivatives of the (1→3)–α-D-glucan which were isolated from the spores of *G. lucidum*. The process included aminopropylated, sulfated, carboxymethylated, carboxymethylated and sulfated, and benzylamidated-carboxymethylated with varying degrees of substitution synthesized. These modified derivatives have shown potent stimulating effects on the lymphocyte proliferation and antibody production. On the other hand, the induction of carboxymethyl group with low degree of substitution was the best choice for the improvement of the immunostimulating activity. A one-step to prepared selenium nanoparticles (SeNPs) decorated by the water–soluble derivative of *G. lucidum* polysaccharides (SPS) was tested on their anti-inflammatory activity against murine RAW264.7 macrophage cells induced by lipopolysaccharides [51]. The results suggested that seNPs-SPS complexes possessed anti-inflammatory potential modulating pro/anti-inflammation cytokine secretion profiles. The isolated polysaccharide (molecular weight 1×10^6 Da) isolated from the sporocarps of *G. lucidum* was used to determined radioprotective property *in vivo* and *in vitro* by survival studies [52]. The findings indicated *Ganoderma* polysaccharides were significantly protective against radiation-induced damages. Another study showed positive effect of polysaccharides extracted from *G. lucidum* on blood glucose, serum level, lipid peroxidation, nonenzymaic and enzymic antioxidants in the plasma and liver of streptozotocin-induced diabetic rats [53]. The neutral crude polysaccharides extracted from *G. lucidum* fruiting body claimed by Chen *et al.* [54] exhibited the higher DPPH-, O- and OH- free radical scavenging activities. It could significantly enhance the antioxidant enzyme activities (SOD, CAT and GPx) and reduce levels of IL-1β, IL-6 and TNF-α in rats with cervical cancer. Two fractions purified from the fruiting body of *G. lucidum* were tested on activation of macrophage cell (RAW 264.7) and antitumor activities on the human breast cancer cell (MDA-MB-231) [55]. The results indicated that both fractions increased the proliferation and pinocytic activity of macrophage significantly and played an inhibiting effect on the cancer cell.

5. Conclusion

The method of *G. lucidum* polysaccharides extraction and fractionation varied greatly according to the objectives of each study. This targeted end products and the yields of polysaccharides extraction normally started with hot water extraction. The temperature applied could be lower than boiling point of water in order to preserve the proteins. Such temperature can soften some fibers and the water molecules in the liquid phase are easily absorbed by the mushroom cells. Therefore, most of the initial steps of polysaccharide extraction need to ensure the physical form of resources are pulverized to obtain the maximum effects on the samples, since the polysaccharides contain many —OH groups which can be easily extracted by the water. Although most of the methodology of the *G. lucidum* polysaccharides extraction and fractionation are different they all followed a similar principle, i.e., using ethanol to precipitate the main polysaccharides. However, at this stage, the extract is still considered crude. Some reported studies would terminate at this stage and use the crude for therapy. In fact, the extraction should continue to the stage of purification in order to separate the different types of polysaccharides in crude. This paper has reviewed six different types of fractionations from different media of *G. lucidum.* All these studies reported positive results in accordance to the study objective(s).

Acknowledgements

The authors thank the Director General of Health, Malaysia for the permission to publish and the Director of Institute for Medical Research (IMR), Kuala Lumpur for the support. This work was financially supported by the Ministry of Health Research Grant Scheme (Grant ID: NMRR-11-953-10736). The authors also thank Prof. Vikineswary (University of Malaya), Dr. Wan Abd Al Qadr Imad Wan mothar (UM) and Dr. Rahayu Ahmad (MIP) for critically reviewing this chapter.

Author details

Yew-Keong Choong[1]*, Kavithambigai Ellan[2], Xiang-Dong Chen[3] and
Shaiful Azuar Mohamad[4]

*Address all correspondence to: yewkeong11@yahoo.co.uk

1 Phytochemistry Unit, Herbal Medicine Research Center, Institute for Medical Research, Jalan Pahang, Kuala Lumpur, Malaysia

2 Virology Unit, Infectious Disease Research Center, Institute for Medical Research, Jalan Pahang, Kuala Lumpur, Malaysia

3 Institute for Medical Research, Jalan Pahang, Kuala Lumpur, Malaysia

4 Malaysia Nuclear Agency, Kajang, Selangor, Malaysia

References

[1] Ikekawa T, Uehara N, Maeda Y, Nakanishi M, Fukuoka F. Antitumor activity of aqueous extracts of edible mushrooms. Cancer Research. 1969;**29**:734-735

[2] Rahar S, Swami G, Nagpal N, Nagpal MA, Singh GS. Preparation, characterization, and biological properties of β-glucans. Journal of Advanced Pharmaceutical Technology & Research. 2011;**2**:94-103

[3] Kobayashi H, Matsunaga K, Oguchi Y. Antimetastatic effects of PSK (Krestin), a protein-bound polysaccharide obtained from basidiomycetes: An overview. Cancer Epidemiology, Biomarkers & Prevention. 1995;**4**(3):275-281

[4] Zeng D, Zhua S. Purification, characterization, antioxidant and anticancer activities of novel polysaccharides extracted from Bachu mushroom. International Journal of Biological Macromolecules. 2018;**107**:1086-1092

[5] Liu M, Li S, Wang X, Zhu Y, Zhang J, Liu H, Jia L. Characterization, anti-oxidation and anti-inflammation of polysaccharides by *Hypsizygus marmoreus* against LPS-induced toxicity on lung. International Journal of Biological Macromolecules. 2018;**111**:121-128

[6] Wang J, Wang C, Li S, Li W, Yuan G, Pan Y, Chen H. Antidiabetic effects of *Inonotus obliquus* polysaccharides in streptozotocininduced type 2 diabetic mice and potential mechanism via PI3K-Akt signal pathway. Biomedicine & Pharmacotherapy. 2017;**95**:669-1677

[7] Yang L, Zhang J, Meng Z. Purification, characterization and anti-tumor activities of polysaccharides extracted from wild *Russula griseocarnosa*. International Journal of Biological Macromolecules. 2018;**109**:1054-1060

[8] Hea X, Wang X, Fang J, Chang Y, Ning N, Guo H, Huang L, Huang X, Zhao Z. Structures, biological activities, and industrial applications of the polysaccharides from *Hericium erinaceus* (Lion's mane) mushroom: A review. International Journal of Biological Macromolecules. 2017;**97**:228-237

[9] Daba AS, Ezeronye OU. Anti-cancer effect of polysaccharides isolated from higher basidiomycetes mushrooms. African Journal of Biotechnology. 2003;**2**(12):672-678

[10] Zhang J, Meng G, Zhai G, Yang Y, Zhao H, Jia L. Extraction, characterization and antioxidant activity of polysaccharides of spent mushroom compost of *Ganoderma lucidum*. International Journal of Biological Macromolecules. 2016;**82**:432-439

[11] Nagaia K, Uenob Y, Tanakab S, Hayashib R, Shinagawaa K, Chayama K. Polysaccharides derived from *Ganoderma lucidum* fungus mycelia ameliorate indomethacin-induced small intestinal injury via induction of GM-CSF from macrophages. Cellular Immunology. 2017;**320**:20-28

[12] Liu Y, Tang Q, Zhang J, Xia Y, Yang Y, Di W, Cui SW. Triple helix conformation of β-D-glucan from *Ganoderma lucidum* and effect of molecular weight on its immunological activity. International Journal of Biological Macromolecules. Available online 13 March 2018

[13] Cao LZ, Lin ZB. Regulatory effect of *Ganoderma lucidum* polysaccharides on cytotoxic T-lymphocytes induced by dendritic cells in vitro. Acta Pharmacologica Sinica. 2003;**24**:321-326

[14] You YH, Lin ZB. Protective effects of *Ganoderma lucidum* polysaccharides peptide on injury of macrophages induced by reactive oxygen species. Acta Pharmacologica Sinica. 2002;**23**:787-791

[15] Giavasis I. Bioactive fungal polysaccharides as potential functional ingredients in food and nutraceuticals. Current Opinion in Biotechnology. 2014;**26**:162-173

[16] Ho YS, Yu MS, Yang XF, So KF, Yuen WH, Chang RC. Neuroprotective effects of polysaccharides from wolfberry, the fruits of *Lycium barbarum*, against homocysteine-induced toxicity in rat cortical neurons. Journal of Alzheimer's Disease. 2010;**19**(3):813-827

[17] Ai-Lati A, Liu S, Ji Z, Zhang H, Mao J. Structure and bioactivities of a polysaccharide isolated from Ganoderma lucidum in submerged fermentation. BIOENGINEERED. 2017;**8**(5):565-571

[18] Guo L, Xie J, Ruan Y, Zhou L, Zhu H, Yun X, Jiang Y, Lü L, Chen K, Min Z, Wen Y, Gu J. Characterization and immunostimulatory activity of a polysaccharide from the spores of *Ganoderma lucidum*. International Immunopharmacology. 2009;**9**:1175-1182

[19] Wang J, Zhang Y, Yuan Y, Yue T. Immunomodulatory of selenium nano-particles decorated by sulphated *Ganoderma lucidum* polysaccharides. Food and Chemical Toxicology. 2014;**68**:183-189

[20] Huang S-Q, Li J-W, Li Y-Q, Wang Z. Purification and structural characterization of a new water-soluble neutral polysaccharide GLP-F1-1 from *Ganoderma lucidum*. International Journal of Biological Macromolecules. 2011;**48**:165-169

[21] Liu W, Wang H, Pang X, Yao W, Gao X. Characterization and antioxidant activity of two low-molecular-weight polysaccharides purified from the fruiting bodies of *Ganoderma lucidum*. International Journal of Biological Macromolecules. 2010 May 1;**46**(4):451-457

[22] Tomoda M, Gonda R, Kasahara Y, Hikino H. Glycan structures of ganoderans b and c, hypoglycemicglycansof*Ganodermalucidum*fruitbodies.Phytochemistry;**25**(12):2817-2820

[23] Ruiz-Herrera J. Fungal Cell Wall: Structure, Synthesis, and Assembly. Boca Raton, Ann Arbor, London: CRC Press; 1956

[24] Mizuno T. Development of antitumor polysaccahrides from mushroom fungi. Food and Food Ingredient Japanese Journal. 1996;**167**:69-87

[25] Pan D, Wang L, Chen C, Teng B, Wang C, Xu Z, Hu B, Zhou P. Structure characterization of a novel neutral polysaccharide isolated from *Ganoderma lucidum* fruiting bodies. Food Chemistry. 2012;**135**:1097-1103

[26] Jiang H, Sun P, He J, Shao P. Rapid purification of polysaccharides using novel radial flow ion-exchange by response surface methodology from *Ganoderma lucidum*. Food and Bioproducts Processing. 2012;**90**:1-8

[27] Dong Q, Wang Y, Shi L, Yao J, Li J, Ma F, Ding K. A novel water-soluble b-D-glucan isolated from the spores of *Ganoderma lucidum*. Carbohydrate Research. 2012;**353**(2012):100-105

[28] Bao XF, Liu CP, Fang JN, Li XY. Carbohydrate Research. 2001;**332**:67-74

[29] http://tcmadaptogen.com/msg.php?id=310

[30] Rajalingam D, Loftis C, Jiashou JX, Kumar TKS. Trichloroacetic acid-induced protein precipitation involves the reversible association of a stable partially structured intermediate. Protein Science. 2009;**18**(5):980-993

[31] http://www.bio-rad.com/en-my/applications-technologies/anion-exchange-chromatography

[32] DuBois M, Gilles KA, Hamilton JK, Rebers PA, Smith F. Colorimetric method for determination of sugars and related substances. Anlytical Chemistry. 1956;**28**:350-356

[33] https://www.gelifesciences.com/shop/chromatography/resins/size-exclusion/sephacryl-s-200-hr-p-05621?current=17058410

[34] Xu S, Dou Y, Ye B, Wu Q, Wang Y, Hu M, Ma F, Rong X, Guo J. *Ganoderma lucidum* polysaccharides improve insulin sensitivity by regulating inflammatory cytokines and gut microbiota composition in mice. Journal of Functional Foods. 2017;**38**(2017):545-552

[35] Xiao C, Wu Q, Zhang J, Xie Y, Cai W, Tan J. Antidiabetic activity of *Ganoderma lucidum* polysaccharides F31 downregulated hepatic glucose regulatory enzymes in diabetic mice. Journal of Ethnopharmacology. 2017;**196**(2017):47-57

[36] https://www.gelifesciences.com/shop/chromatography/resins/size-exclusion/sephacryl-s-300-hr-p-05620?current=17059901

[37] Liu Y, Zhang J, Tang Q, Yang Y, Guo Q, Wang Q, Wu D, Cui SW. Physicochemical characterization of a high molecular weight bioactive-β-glucan from the fruiting bodies of *Ganoderma lucidum*. Carbohydrate Polymers. 2014, January 2014;**101**(30):968-974

[38] https://en.wikipedia.org/wiki/Trifluoroacetic_acid

[39] https://www.quora.com/Which-one-is-more-acidic-trichloroacetic-acid-or-trifluoroacetic-acid

[40] https://en.wikipedia.org/wiki/Hydrolysis

[41] Khan I, Huang G, Li X, Leong W, Xia W, Wendy Hsiao WL. Mushroom polysaccharides from *Ganoderma lucidum* and *Poria cocos* reveal prebiotic functions. Journal of Functional Foods. 2018;**41**:191-201

[42] Li K, Yu M, Hu Y, Ren G, Zang T, Xu X, Qu J. Three kinds of *Ganoderma lucidum* polysaccharides attenuate DDC-induced chronic pancreatitis in mice. Chemico-Biological Interactions. 2016;**247**:30-38

[43] Sanodiya BS, Thakur GS, Baghel RK, Prasad GB, Bisen PS. Ganoderma lucidum: A potent pharmacological macrofungus. Current Pharmaceutical Biotechnology. 2009;**10**(8): 717-742

[44] Boh B, Berovic M, Zhang J, Zhi-Bin L. *Ganoderma lucidum* and its pharmaceutically active compounds. Biotechnology Annual Review. 2007;**13**:265-301

[45] Yang FC, Liau CB. The influence of environmental conditions on polysaccharide formation by *Ganoderma lucidum* in submerged cultures. Process Biochemistry. 1998;**33**:547-553

[46] Mau JL, Lin HC, Chen CC. Non-volatile components of several medicinal mushrooms. Food Research International. 2001;**34**:521-526

[47] Li X, Zhao R, Zhou HL, Wu DH. Deproteinization of polysaccharide from the stigma maydis by sevag method. Advanced Materials Research. 2012;**340**:416-420

[48] Borchers AT, Stern JS, Hackman RM, Keen CL, Gershwin ME. Minireview: Mushrooms, tumors and immunity. Proceedings of the Society for Experimental Biology and Medicine. 1999;**221**:281-293

[49] http://www.chembio.uoguelph.ca/educmat/chm357/sephadex.pdf

[50] Bao XF, Duan J, Fang X, Fang J. Chemical modifications of the (1→3)-α-D-glucan from spores of *Ganoderma lucidum* and investigation of their physiochemical properties and immunological activity. Carbohydrate Research. 2001;**336**:127-140

[51] Wang J, Zhang Y, Yuan Y, Yue T. Immunomodulatory of selenium nano-particles decoreated by sulfated *Ganoderma lucidum* polysaccharides. Food and Chemical Toxicology. 2014;**68**:183-189

[52] Pillai TG, Nair CKK, Janardhanan KK. Polysaccharides isolated from *Ganoderma lucidum* occurring in southern part of India, protects radiation induced damages both in vitro and in vivo. Environmental Toxicology and Pharmacology. 2008;**26**:80-85

[53] Jia J, Zhang X, Hu Y-S, Wu Y, Wang Q-Z, Li N-N, Guo Q-C, Dong X-C. Evaluation of in vivo antioxidant activities of*Ganoderma lucidum* polysaccharides in STZ-diabetic rats. Food Chemistry. 2009;**115**(1):32-36

[54] Ping CX, Yan C, Bing LS, Guo CY, Yun LJ, Ping LL. Free radical scavenging of *Ganoderma lucidum* polysaccharides and its effect antioxidant enzymes and immunity activities in cervical carcinoma rats. Carbohydrate Polymers. 2009;**77**:389-393

[55] Zhao L, Dong Y, Chen G, Hu Q. Extraction, purification, characterization and antitumor activity of polysaccharides from *Ganoderma lucidum*. Carbohydrate Polymers. 2010;**80**:783-789

Chapter 4

Fractional Distillation of Bio-Oil Produced by Pyrolysis of Açaí (*Euterpe oleracea*) Seeds

Douglas Alberto Rocha de Castro,
Haroldo Jorge da Silva Ribeiro,
Caio Campos Ferreira, Márcio de Andrade Cordeiro,
Lauro Henrique Hamoy Guerreiro,
Anderson M. Pereira, W. G. dos Santos,
Marcelo Costa Santos, Fernanda B. de Carvalho,
Jose Otavio Carrera Silva Junior, R. Lopes e Oliveira,
Sergio Duvoisin, Luiz Eduardo Pizarro Borges and
Nélio Teixeira Machado

Additional information is available at the end of the chapter

http://dx.doi.org/10.5772/intechopen.79546

Abstract

In this work, the seeds of açaí (*Euterpe oleracea*, Mart), a rich lignin-cellulose residue, has been submitted to pyrolysis to produce a bio-oil-like fossil fuels. The pyrolysis carried out in a reactor of 143 L, 450°C, and 1.0 atm. The morphology of Açaí seeds *in nature* and after pyrolysis is characterized by SEM, EDX, and XRD. The experiments show that bio-oil, gas, and coke yields were 4.38, 30.56, and 35.67% (wt.), respectively. The bio-oil characterized by AOCS, ASTM, and ABNT/NBR methods for density, kinematic viscosity, and acid value. The bio-oil density, viscosity, and acid value were 1.0468 g/cm^3, 68.34 mm^2/s, and 70.26 KOH/g, respectively. The chemical composition and chemical functions of bio-oil are determined by GC-MS and FT-IR. The GC-MS identified in bio-oil 21.52% (wt.) hydrocarbons and 78.48% (wt.) oxygenates (4.06% esters, 8.52% carboxylic acids, 3.53% ketones, 35.16% phenols, 20.52% cresols, 5.75% furans, and 0.91% (wt.) aldehydes), making it possible to apply fractional distillation to obtain fossil fuel-like fractions rich in hydrocarbons. The distillation of bio-oil is carried out in a laboratory-scale column, according to the boiling temperature of fossil fuels. The distillation of bio-oil yielded fossil fuel-like fractions (gasoline, kerosene, and light diesel) of 4.70, 28.21, and 22.35% (wt.), respectively.

Keywords: açaí, seeds, pyrolysis, bio-oil, distillation, biofuels

IntechOpen

1. Introduction

During the processing of açaí juice from açaí (*Euterpe oleracea,* Mart) seeds *in nature*, a native palm of natural occurrence in the Amazon region, belonging to the family Arecaceae and compassing approximately 200 genera and about 2600 species, distributed predominantly in tropical and subtropical areas [1], a by-product is produced and/or discharged, the açaí seeds, posing a huge environmental problem of solid waste management in Belém metropolitan region, as well as in the municipalities around the city of Belém-Pará-Brazil.

The State of Pará is the largest national producer of açaí with 1,012,740 ton/year of fruits [2], being the production due to extractive 198,149 tons/year of fruits in the crop year 2014 [3], representing 55.4% (wt.) of the national production of extractive açaí in the crop year 2014, and the production due to agricultural systems using a planted area of 154,500 hectare, was 814,590 tons in the year 2014. Of the total 1,012,740 tons/year of fruits, 8,405,742 tons/year is a residue (açaí seeds) representing approximately 83% (wt.).

The metropolitan region of Belém-Pará-Brazil, capital the State of Pará, has approximately 10,000 stores of açaí commercialization, producing an average of 200 kg açaí seeds/day per store, thus producing around 2000 tons residue/day [4]. In 2015, there was a growth of 27.35% (wt.) in production, 10.86% (wt.) in planted area and 14.88% (wt.) on the specific production yield, compared to 2014 [2]. The seed of açaí is an oil-fiber seed, and according to the literature, constituted by a small solid endosperm attached to a tegument, rich in cellulose with approximately 53.20% (wt.), hemicelluloses 12.26% (wt.), lignin 22.30% (wt.), as well as 3.50% lipids (wt.) [5–9].

In a scenery, the modern industrial society focuses on minimization of global warming and CO_2 emission, as well as energy efficient supply systems and less consumption of fossil-based fuels. To achieve this, the use of renewable energy resources is essential [10]. In this context, processes that minimize the industrial and agro-industrial residues either by reusing or recycling them result in energetic and environmental benefits to the global society. In addition, recycling industrial and agro-industrial residues enables to use raw materials of low cost, making it possible to increase the economic viability of biofuels' production [11].

Among the most important renewable energy sources, this biomass is considered as an important one, since it could be a suitable alternative for conventional fossil fuels [12]. In addition, biomass energy producing systems may be implemented not only close to industrial and agro-industrial production systems, but also in any location where vegetable species can be grown and/or domestic animals are reared [12]. The systematic use of biomass makes it possible to reduce global warming compared to fossil fuel energy systems, as all the vegetable species use and store CO_2 for the photosynthesis process [13]. CO_2 stored in the plant is released when the plant material is burned and/or decays [12, 13]. Thus, by replanting the crops, the new growing vegetable species can use the CO_2 produced by burning vegetable species, as in the carbonization processes (e.g., pyrolysis), and hence contributing to close the carbon dioxide cycle, as reported in the literature by Kelli et al. [14].

The residual açaí seeds, an oil-fiber seed rich lignin-cellulose material, whose centesimal composition reported in the literature is constituted of lipids between 1.65 and 3.56% (wt.), total fibers between 29.69 and 62.75% (wt.), hemicellulose between 9.01 and 14.19% (wt.), cellulose between 39.83 and 40.29% (wt.), lignin between 4.00 and 8.93% (wt.), ash between 0.15 and 1.68% (wt.), moisture between 10.15 and 39.39% (wt.), protein between 5.02 and 7.85% (wt.), 0.83% (wt.) fixed carbon, and 7.82% (wt.) volatile matter approximately [5–9]. A process that makes it possible for the use of açaí seeds, an oil-fiber seed rich lignin-cellulose-based material, of low quality for producing liquid and gaseous fuels is pyrolysis [15, 16].

In the last years, several process schemes have been proposed to remove oxygenate compounds from biomass-derived bio-oils, including molecular distillation to separate water and carboxylic acids from pyrolysis bio-oils [17–19], fractional distillation to isolate chemicals and improve the quality of bio-oil [20–25], liquid-liquid extraction using organic solvents and water to recover oxygenate compounds of bio-oils [26, 27]. Non-conventional separation methods using aqueous salt solutions for phase separation of bio-oils have been also applied [28]. Recently, the bio-oil obtained by pyrolysis of açaí seeds in nature have been upgraded by fractional distillation, as described in detail as follows [15, 16]. Guerreiro et al. [15, 16] investigated the influence of column height by fractional distillation of bio-oil obtained by pyrolysis of açaí seeds at 350°C in pilot scale using Vigreux columns of 10 and 30 cm. The yields of gasoline were 6.60 and 7.12% (wt.), while that of kerosene were 11.05 and 12.64 (wt.), respectively, for columns of 10 and 30 cm, showing no significant variation. The acid value of gasoline-like fraction using Vigreux column of 10 and 30 cm were 17.08 and 16.79 mg KOH/g, respectively, while that of kerosene were 62.34 and 59.35 mg KOH/g, respectively, showing no significant variation. In addition, the kinematic viscosity of gasoline-like fraction using columns of 10 and 30 cm were 1.58 and 1.45 mm^2/s, respectively, while that of kerosene were 4.04 and 3.10 mm^2/s, a variation between 8.23 and 23.27% showing that kinematic viscosity is more sensitive to the influence of column height, decreasing with column height.

In this work, the pyrolysis of Açaí seeds (*Euterpe oleracea*, Mart) has been systematically investigated in pilot scale at 450°C 1.0 atmosphere to produce a bio-oil, a pyrolysis reaction liquid product, been submitted to fraction distillation carried out in a laboratory-scale column (Vigreux Column) according to the boiling temperature range of fossil fuels to study the feasibility of producing fossil fuels like fractions (gasoline, kerosene, and diesel), as well as the morphology of solid phase products (coke), of açaí seeds (*Euterpe oleracea*, Mart) pyrolysis process at 450°C.

2. Materials and methods

2.1. Materials

The seeds of Açaí (*Euterpe oleracea*, Mart) obtained in a small Store of Açaí Commercialization, located in the District of Guamã, Belém-Pará-Brazil. **Figure 1** shows the anatomy of aça í fruits (cross section): (1) Embryo, (2) Endocarp, (3) Scar, (4) Pulp, (5) Pericarp + Tegument, and (6) Mesocarp.

2.2. Pre-treatment of açaí (*Euterpe oleracea*, Mart) seeds

The seeds of Açaí (*Euterpe oleracea*, Mart) are submitted to drying at 105°C using a pilot oven with air recirculation (SOC, FABBE, Ltd, Brazil, Model: 170) for a period of 24 h. Afterward,

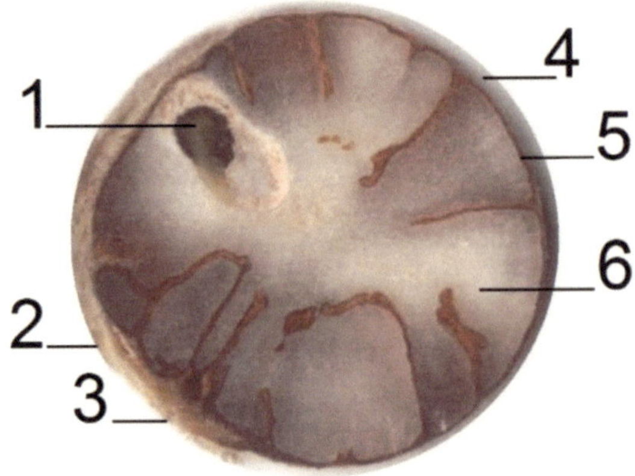

Figure 1. Anatomy of Açaí (*Euterpe oleracea,* Mart) fruit *in nature* (cross section): (1) embryo, (2) endocarp, (3) scar, (4) pulp, (5) pericarp + tegument, and (6) mesocarp.

the dried seeds are grinded using a laboratory knife cutting mill (TRAPP, Brazil, Model: TRF 600). Then, the dried and grinded açaí seeds are sieved using an 18 Mesh sieve to remove the excess fiber material.

2.2.1. Centesimal and elemental characterization of açaí (Euterpe oleracea, Mart) seeds

The centesimal and elemental characterizations of açaí (*Euterpe oleracea,* Mart) seeds are performed for moisture (AOAC 935.29), volatile matter (ASTM D 3175-07), ash (ASTM D 3174-04), fixed carbon (ASTM D6316-09), lipids (AOAC 963.15), proteins (AOAC 991.20), fibers according to the official methods reported in the literature [29], and insoluble lignin according to the method of Klason described elsewhere [30].

2.3. Fractional distillation of bio-oil

2.3.1. Distillation: experimental apparatus and procedures

The fractional distillation of bio-oil is performed by using an experimental apparatus, as described in the literature [31]. The distillation apparatus, illustrated in **Figure 2**, has an electrical heating blanket of 480 W (Fisaton, Model: 202E, Class: 300), which is thermostatically controlled, a 500 ml round bottom, and two neck flasks with outer joints. The side joint is used to insert a long-thin thermocouple of a digital thermometer, and the center joint is connected to a distillation column (Vigreux) of 30 cm. The center top-outer joint is connected to the bottom inner joint of a Liebig glass-borosilicate condenser. The Liebig glass-borosilicate condenser is connected to a 250 ml glass separator funnel by the top-outer joint. A thermocouple connected to the top-outer joint 24/40 of the distillation column makes it possible to measure the vapor temperature at the top of the borosilicate-glass distillation columns (Vigreux). A cryostat bath provides cold water at 15°C to the Liebig glass-borosilicate condenser. The 500-ml round-bottom borosilicate-glass flask and the distillation column are insulated with glass wool and aluminum foil sheet to avoid heat losses, respectively. The mass of distillation

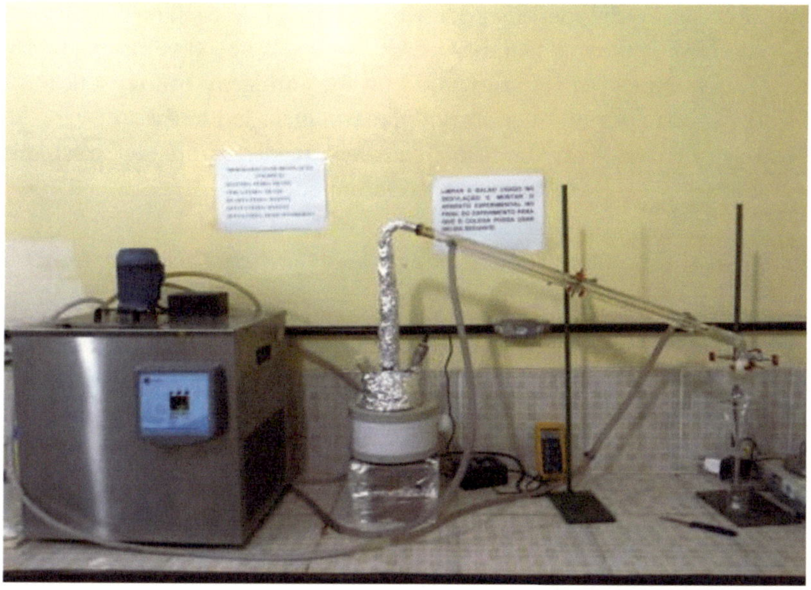

Figure 2. Vigreux borosilicate-glass distillation column of 500 ml, electrical heating mantel, cryostat bath, Liebig condenser, and separating funnel.

fractions (gasoline, kerosene, and light diesel-like fuels) is recorded and weighed. The distillation fractions are submitted to the pre-treatment of decantation to separate the aqueous and organic phases.

2.4. Physicochemical analysis of bio-oil and distillation fractions and chemical composition of bio-oil

2.4.1. *Physicochemical analysis of bio-oil and distillation fractions*

Bio-oil and the distillation fractions are obtained according to the boiling temperature range of fossil fuels (gasoline, kerosene, and diesel) physicochemical characterized for acid value (AOCS Cd 3d-63), density (ASTM D4052) at 25°C, kinematic viscosity (ASTM D445/D446), and refractive index (AOCS Cc 7-25), as described in the literature [31]. The qualitative analyses of chemical functions present in the bio-oil are performed by FT-IR spectroscopy, according to the literature [31].

2.4.2. *GC-MS of bio-oil*

The separation and identification of all the compounds present in bio-oil are performed by GC-MS, using a gas chromatograph (Agilent Technologies, USA, Model: GC-7890B) coupled to MS-5977A Mass Spectrometer, a SLBTM-5 ms (30 m × 0.25 mm × 0.25 mm) fused silica capillary column. The temperature conditions used in the GC-MS were: injector temperature: 250°C; split: 1:50, detector temperature: 230°C and quadrupole: 150°C; injection volume: 1.0 mL; oven: 60°C/1 min; 3°C/min; 200°C/2 min; 20°C/min; 230°C/10 min. The intensity, retention time, and compound identification were recorded for each peak analyzed according to the NIST (Standard Reference Database 1A, V14) mass spectra library, which is part of the software. The identification is made based on the similarity of the peak mass spectrum obtained with the spectra within the library database, included in the software. The contents of all identified oxygenates and hydrocarbons present in each sample were separated and the chemical composition of each experiment was estimated.

2.5. Morphology of solid phase products of açaí seeds (*Euterpe oleracea*, Mart)

The characterization of solid phase products (coke) obtained by pyrolysis of dried Açaí (*Euterpe oleracea*, Mart) seeds at 450°C and 1.0 atm, in pilot scale, was performed by scanning electron microscope (SEM), energy dispersive X-ray spectroscopy (EDX), and X-ray diffraction (XRD) according the equipment's and procedures described in detail elsewhere [31, 32].

3. Results and discussions

3.1. Centesimal and elemental characterization of açaí (*Euterpe oleracea*) seeds

Table 1 shows the centesimal and elemental characterization of açaí (*Euterpe oleracea*) seeds *in nature*. The experimental results obtained for moisture, proteins, and cellulose are according to those reported by Altman [9], while those for lipids, proteins, and fibers are according to those reported by Kabacknik and Roger [8]. In addition, the results for lipids and proteins are according to those reported by Tamiris et al. [7]. The centesimal characterization of açaí (*Euterpe oleracea*) seeds totalizes 98.07% (wt.) in dry basis, showing that summation (moisture, lipids, proteins, fibers, hemicellulose, cellulose, lignin, volatile matter, fixed carbon, and ash) is almost close to 100% (wt.). The summation of centesimal characterization of açaí seeds (moisture, lipids, proteins, fibers, ash, and nitrogen) reported by Tamiris et al. [7] is also close to 100% (wt.). The results for fibers are much higher than those reported by Tamiris et al. [7], who reported 85.69% (wt.) of carbohydrates, and much lower than those reported by Altman [9].

3.2. Process parameters and overall steady state material balances of dried açaí (*Euterpe oleracea*, Mart) seeds pyrolysis

The process conditions and steady state material balances of dried Açaí (*Euterpe oleracea*, Mart) seeds pyrolysis are shown in **Table 2**. The experimental results show that bio-oil, gas, water phase, and coke yields were 4.38, 29.39, and 35.67% (wt.), respectively. The bio-oil yield of 4.39% (wt.) is lower compared to similar data for bio-oil yield obtained by fast pyrolysis of forestry residues at 520°C reported in the literature [33, 34], ranging from 10 to 20% (wt.), and depends on the feedstock composition. The low bio-oil yield is probably due to the high fiber content, as illustrated in **Table 3**. The high yield of water phase is probably due to dehydration reactions along the pyrolysis process, as the initial moisture content is 10.15% (wt.), being the water phase yield of 29.39% (wt.) close to that of 28.0% (wt.), as reported in the literature [34].

Physicochemical analysis	Cordeiro [6]	Tamiris et al. [7]	Kabacknik and Roger [8]	Altman [9]
	Wet Basis	Dry Basis	Wet Basis	Wet Basis
Moisture (%)	10.15	0.79	58.30	13.60
Lipids (%)	0.61	1.98	1.65	3.48
Proteins (%)	6.25	7.85	5.56	5.02
Fibers (%)	29.79	2.1	21.29	62.95
Hemicelluloses (%)	5.5	—	—	14.19
Cellulose (%)	40.29	—	—	39.83
Lignin (%)	4.00	—	—	8.93
Volatile matter (%)	0.5	—	—	—
Fixed carbon (%)	0.83	—	—	—
Ash (%)	0.15	1.68	5.97	1.55
Nitrogen (%)	—	1.26	—	—
Carbohydrate (%)	—	85.69	—	—

Table 1. Centesimal and elemental characterization of Açaí (*Euterpe oleracea*, Mart) seeds *in nature*.

3.3. Physicochemical characterization of bio-oil

Table 3 presents the physicochemical characterization of bio-oil obtained by pyrolysis of dried açaí (*Euterpe oleracea*, Mart) seeds at 450°C and 1.0 atm. The bio-oil density and viscosity were 1.0468 g/cm^3 and 68.34 mm^2/s, respectively. The density and kinematic viscosity are according to similar data reported in the literature [35], where average density of wood bio-oil is 1.2 g/cm^3 and the kinematic viscosity of wood bio-oils at 50°C varies between 40 and 100 mm^2/s. The acid value of bio-oil was 70.26 KOH/g, being the acidity due to the presence of oxygenates compounds, such as carboxylic acids, phenols, cresols, ketones, and aldehydes, as described in **Table 5**, confirming the results reported by Oasmaa et al. [36], who stated that acidity of fast pyrolysis bio-oil is mainly due to volatile carboxylic acids, but not only, as well as other functional groups such as phenols, resin acids, and hydroxy acids [36].

3.4. Mass balances, yields (distillates and raffinate) of fractional distillation, and physicochemical characterization of distillation fractions of bio-oil obtained by pyrolysis of dried açaí (*Euterpe oleracea*, Mart) seeds

Mass balances and yields (distillates and raffinate) of fractional distillation of bio-oil obtained by pyrolysis of dried Açaí (*Euterpe oleracea*, Mart) seeds at 450°C and 1.0 atm are illustrated in **Table 4**. The distillation of bio-oil yielded fossil fuel-like fractions (gasoline, kerosene, and light diesel) of 4.70, 28.21, and 22.35% (wt.), respectively, totalizing 55.26% (wt.), being according to similar results for distillation of bio-oil reported in the literature [22, 23, 25, 37]. Zheng and Wie [22], investigated the distillation of fast pyrolysis bio-oil at 80°C and 15 mmHg, obtaining a distilled bio-oil yield of 61% (wt.), being the oxygenates' content of distilled bio-oil 9.2% (wt.).

Process parameters	Cracking temperature (°C)
	450
Mass of Açaí (kg)	30
Mass of GLP (kg)	
Cracking time (min)	150
Mechanical stirrer speed (rpm)	0
Initial cracking temperature (°C)	179
Mass of aqueous phase (OLP + H_2O) (kg)	10,133
Mass of Coke (kg)	10,700
Mass of OLP (kg)	1316
Mass of H_2O (kg)	8.816
Mass of gas (kg)	9167
Yield of OLP (kg)	4.39
Yield of coke (kg)	35.67
Yield of H_2O (kg)	29.39
Yield of Gas (kg)	30.56

Table 2. Process parameters and overall steady state material balances of dried Açaí (*Euterpe oleracea*, Mart) seeds pyrolysis at 450°C and 1.0 atm, in pilot scale.

Physicochemical properties	450°C	ANP N° 65
	Bio-oil	
ρ (g/cm³)	1.0468	0.82–0.85
Acid value (mg KOH/g)	70.26	—
Refractive index (—)	ND	—
ν (cSt)	68.34	2.0–4.5

ANP: Brazilian National Petroleum Agency, resolution N° 65 (specification of diesel S10).

Table 3. Physicochemical characterization of bio-oil obtained by pyrolysis of dried Açaí (*Euterpe oleracea*, Mart) seeds at 450°C and 1.0 atm, in pilot scale.

Zhang et al. [23] investigated the atmospheric distillation of fast pyrolysis bio-oil and reported an accumulated distillate of 51.86% (wt.). The major organic compounds identified in distillate fractions include phenols, guaiacols, furans, and volatile carboxylic acids (acetic acid and propanoic acid) were also observed in raw bio-oil [23]. In addition, Zhang et al. [23] reported that as the distillation temperature reached 240°C, condensation reactions take place, generating water. Elkasabi et al. [25] reported organic yields from distillation of tail-gas reactive pyrolysis (TGRP) bio-oil ranging from 55 to 65% (wt.).

Distillation: vigreux column	OLP	Gas	Raffinate	Distillates (g)					Yield (wt.%)				
	(g)	(g)	(g)	H_2O	G	K	LD	HD	H_2O	G	K	LD	HD
450°C	136.84	0	40.98	20.26	6.43	38.60	30.59	0	14.80	4.70	28.21	22.35	0

G = gasoline, K = kerosene, LD = light diesel, HD = heavy diesel.

Table 4. Mass balances and yields (Distillates and Raffinate) of fractional distillation of bio-oil obtained by pyrolysis of dried Açaí (*Euterpe oleracea,* Mart) seeds at 450°C and 1.0 atm, in pilot scale.

3.4.1. Physicochemical characterization of distillation fractions (gasoline, kerosene, and light diesel-like fractions)

The physicochemical characterization of distillation fractions (gasoline: 80–175°C, kerosene: 175–235°C, and light diesel-like fraction: 235–305°C) of bio-oil obtained by pyrolysis of dried Açaí (*Euterpe oleracea,* Mart) seeds at 450°C and 1.0 atm is shown in **Table 5**. It can be observed that acidity of distillation fractions (gasoline, kerosene, and light diesel-like) increases with increasing boiling temperature, showing a drastic decrease compared to the acidity of raw bio-oil. This is probably due to the high concentration of higher boiling-point compounds in the distillate fractions, such as p-cresol, o-cresol, guaiacol, phenol, and furans, which increases with the increasing boiling temperature [23]. In addition, the densities and viscosities of kerosene and light diesel-like fractions increase with increasing boiling temperature.

3.5. Qualitative and compositional analyses of bio-oil

3.5.1. Qualitative analyses of chemical functions in bio-oil by FT-IR spectroscopy

Figure 3 illustrates the FT-IR analysis of bio-oil obtained by pyrolysis of dried Açaí (*Euterpe oleracea,* Mart) seeds at 450°C and 1.0 atm, in pilot scale. The identification of absorption bands/peaks done according to previous studies [31, 38, 39]. The spectrum of bio-oil presents a wide band of axial deformation at 3435 cm^{-1}, characteristic of O–H intramolecular hydrogen bond, indicating probably the presence of carboxylic acids. The spectra of bio-oil exhibit intense peaks between 2921 and 2964 cm^{-1} and between 2858 and 2964 cm^{-1}, indicating the presence of aliphatic compounds, associated to methylene (CH_2) and methyl (CH_3) groups. This confirms the presence of hydrocarbons [31, 38]. It has been observed that for bio-oil, an intense axial deformation band, characteristic of carbonyl (C=O) groups, with the peaks at 1742, 1745, and 1747 cm^{-1}, probably associated to a ketone and/or carboxylic acids [31, 38]. The spectra of bio-oil exhibit between 1455 and 1465 cm^{-1}, a characteristic asymmetrical deformation vibration of methylene (CH_2) and methyl (CH_3) groups, indicating the presence of alkanes [31, 38]. The spectrum of bio-oil identified at 1377 cm^{-1}, a band of symmetrical angular deformation of C–H bonds in methyl group (CH_3) [31, 38]. The peaks between 995 and 905 cm^{-1} for bio-oil, are characteristic of an angular deformation outside the plane of C–H bonds, indicating the presence of alkenes [31, 38]. The spectra of bio-oil exhibit bands between 721 and 667 cm^{-1}, peaks characteristic of an angular deformation outside the plane of C–H bonds in methylene (CH_2) group, indicating the presence of olefins [31, 38]. The characteristic peaks of phenols

Physicochemical properties	450° C			ANP N° 65
	G	K	LD	
ρ (g/cm³)	SNA	0.9191	0.9816	**0.82–0.85**
I.A (mg KOH/g)	19.94	61.08	64.78	
I.R (–)	1455	1479	1497	
ν (cSt)	SNA	4.29	9.05	**2.0–4.5**

I.A = acid value, I.R = refractive index, SNA = amount of sample not enough for analysis.

Table 5. Physicochemical characterization of distillation fractions (gasoline: 80–175°C, kerosene: 175–235°C, and light diesel-like fraction: 235–305°C) of bio-oil obtained by pyrolysis of dried Açaí (*Euterpe oleracea*, Mart) seeds at 450°C and 1.0 atm, in pilot scale.

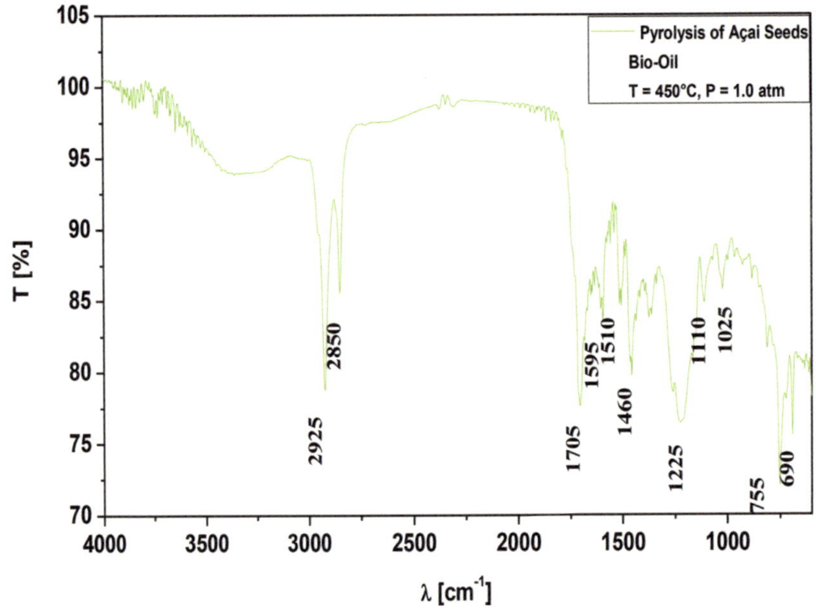

Figure 3. FT-IR of Açaí (*Euterpe oleracea*, Mart) seeds bio-oil after pyrolysis at 450°C and 1.0 atm, in pilot scale.

at 1510 cm^{-1} corresponded to the C=C aromatic ring vibrations [39]. The peaks at 1240 and 1180 cm^{-1} corresponded to the C–C–O asymmetric stretch and C–H in-plane deformations, respectively, while the 990 and 747 cm^{-1} peaks belonged to the C–H out-of-plane vibrations. The frequency due to OH in-plane bonding vibration in phenols, in general, lies in the region 1150–1250 cm^{-1}. The 1500 cm^{-1} vibration is a triplet appearing at 1515 and 1460 cm^{-1}, corresponding probably to the presence of p-cresol and m-cresol, respectively. The OH deformation and C–O stretching vibrations in phenols are close to each other, and therefore they are strongly coupled. They fall above 1100 $cm^{-1}$and extend up to 1330 cm^{-1}. A broad absorption is observed in this region due to the presence of numerous phenols. The out-of-plane hydrogen vibrations appearing in the region 900–675 cm^{-1} suggest the presence of m-cresol and p-cresol. The peaks appearing in the range of 1000–1200 cm^{-1} indicate the presence of C–O–C bond, associated with those in a lower range of 650–750 cm^{-1}, from –CH=CH– bonds, showing the presence of furans, coupled with peaks in the 3000–3100 and 1400–1600 cm^{-1}, suggesting the presence of aromatic rings in the form of C–H and C=C stretching, respectively, corresponding to the presence of furans (benzofuran) [39]. The FT-IR analysis of bio-oil identifies the presence of hydrocarbons (alkanes, alkenes, aromatic hydrocarbons, etc.,) and oxygenates (phenols, cresols, carboxylic acids, ketones, furans, etc.).

3.5.2. *Compositional analyses of bio-oil by GC-MS*

Figure 4 illustrates the chromatogram of bio-oil obtained by pyrolysis of dried Açaí (*Euterpe oleracea*, Mart) seeds at 450°C and 1.0 atm, in pilot scale. The classes of compounds, summation of peak areas, CAS numbers, and retention times of chemical compounds are identified by GC-MS described in **Table 6.** The chemical compounds identified by GC-MS were hydrocarbons (alkanes, alkenes, aromatic hydrocarbons, and cycloalkenes) and oxygenates (esters, phenols, cresols, carboxylic acids, ketones, furans, and aldehydes). The bio-oil is composed

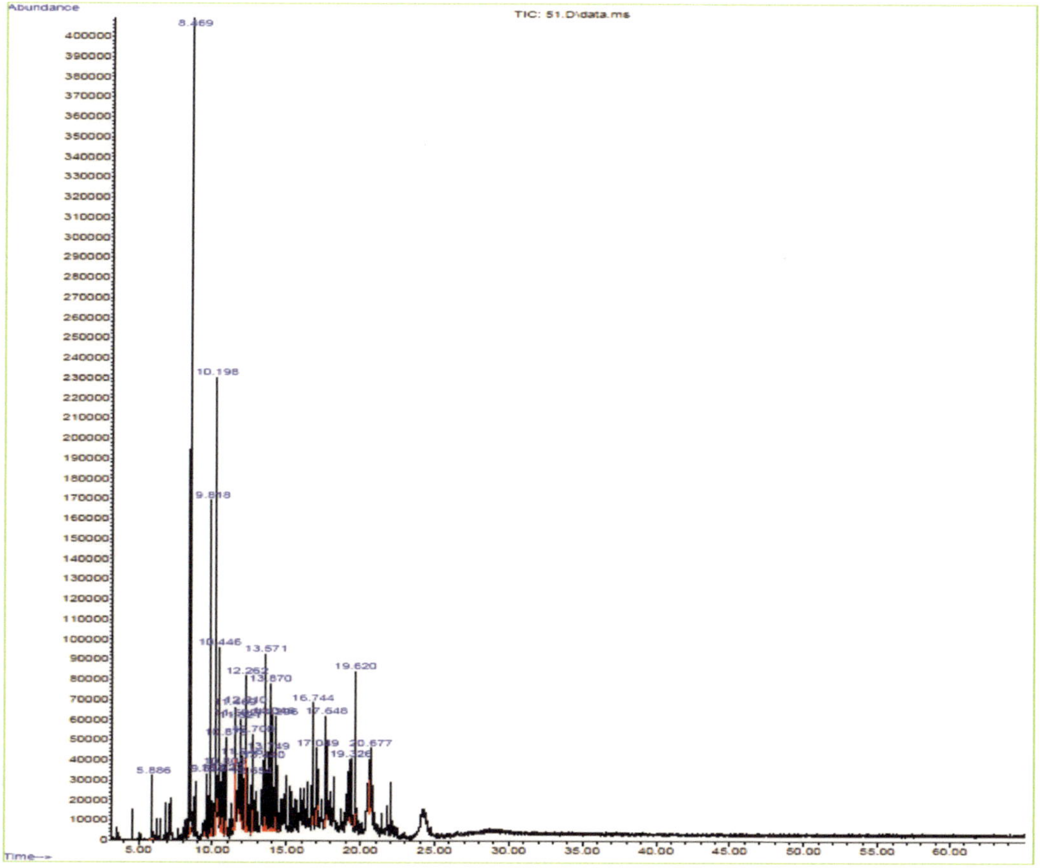

Figure 4. GC-MS of bio-oil obtained by pyrolysis of Açaí (*Euterpe oleracea*, Mart) seeds at 450°C and 1.0 atm, in pilot scale.

450°C			
Class of compounds: chemical compounds	**RT (min)**	**CAS**	**ω_i% (Area)**
Alkanes			
Undecane	10,622	1120–21–4	1124
Tridecane	13,870	629–50–5	2481
Pentadecane	16,744	629–62–9	2290
Dodecane, 5,8-diethyl	19,326	24.251–86–3	1626
Σ (Area.%) =			**7521**
Alkenes			
6-Tridecene, (Z)-	1626	6508–77–6	2118
Σ (Area.%) =			**2118**
Cycloalkenes			
Megastigma-4,6(E), 8 (Z)-trien	13,440	5298–13–5	1847
Σ (Area.%) =			**1847**
Aromatic hydrocarbons			
Naphthalene	12,262	91–20–3	4399
Naphthalene, 1-methyl	14,046	90–12–0	2390
1H-Indene, 1-ethylidene	14,296	2471–83–2	3249
Σ (Area.%) =			**10,038**
Esters			
Undecanoic acid, 10-methyl-, methyl ester	17,049	5129–56–6	1096
Methyl tetradecanoate	19,620	124–10–7	2969
Σ (Area.%) =			**4065**
Carboxylic acids			
Dodecanoic acid	17,648	334–48–5	4307
Tetradecanoic acid	20,677	544–63–8	4216
Σ (Area.%) =			**8523**
Ketones			
2-Pentanone, 4-hydroxy-4-methyl	5886	123–42–2	1878
2-Cyclopenten-1-one, 2,3-dimethyl	9552	1121–05–7	1655
Σ (Area.%) =			**3533**
Phenols			
Phenol	8469	108–95–2	15,932
Phenol, 2-methoxy	10,446	90–05–1	4583

450°C			
Class of compounds: chemical compounds	**RT (min)**	**CAS**	**ω_i% (Area)**
Phenol, 2,6-dimethyl	10,805	576–26–1	1991
Phenol, 2,4-dimethyl	11,469	105–67–9	2034
Phenol, 2,5-dimethyl	11,502	95–87–4	2215
Phenol, 3,4-dimethyl	11,821	95–65–8	3845
Phenol, 4-ethyl-2-methoxy	13,571	2785–89–9	4567
Σ (Area.%) =			**35,167**
Cresols			
p-Cresol	9818	108–39–4	6331
m-Cresol	10,198	106–44–5	11,054
Cresol	12,210	93–51–3	3141
Σ (Area.%) =			**20,526**
Furans			
Benzofuran, 2-methyl	10,879	4265–26–2	1879
Furan, 2-(2 furanylmethyl)-5-methyl	11,946	13,678–51–8	2089
Benzofuran, 4,7-dimethyl	12,700	28,715–26–6	1783
Σ (Area.%) =			**5751**
Aldehydes			
Cinnamaldehyde, β-methyl-	12,654	1196–67–4	0.910
Σ (Area.%) =			**0.910**

Table 6. Classes of compounds, summation of peak areas, CAS numbers, and retention times of chemical compounds identified by GC-MS of bio-oil obtained by pyrolysis of dried Açaí (*Euterpe oleracea,* Mart) seeds at 450°C and 1.0 atm, in pilot scale.

by 21.52% (area) hydrocarbons (2.12% alkenes, 7.52% alkanes, 10.04% aromatic hydrocarbons, and 1.85% cycloalkenes), and 78.48% (area) oxygenates (4.06% esters, 8.52% carboxylic acids, 3.53% ketones, 35.16% phenols, 20.52% cresols, 5.75% furans, and 0.91% aldehydes). The presence of carboxylic acids, as well as phenols and cresols, but not only, confers the high acidity of bio-oil, as described in **Table 3**.

The chemical composition of bio-oil is according to similar bio-oil compositions reported in the literature [22, 23, 25, 26, 36, 37], showing the presence of hydrocarbons, phenols, cresols, furans, carboxylic acids, esters, among other classes of chemical compounds. The hydrocarbons identified in bio-oil by GC-MS present carbon chain length between C_{11} and C_{15} with following carbon chain lengths, alkenes C_{13}, alkanes C_{11}–C_{15}, and cycloalkenes C_{13}. The chemical composition of bio-oil indicates the presence of heavy gasoline compounds with C_{11} (C_5–C_{11}), kerosene-like fractions (C_{11}–C_{12}), and light diesel-like fractions (C_{13}–C_{15}), as observed by fractional distillation illustrated in **Table 4**.

3.6. Morphology of solid phase products of açaí seeds (*Euterpe oleracea*, Mart)

3.6.1. SEM analysis of solid phase

The scanning electron microscopies of açaí (*Euterpe oleracea,* Mart) seeds *in nature* and after pyrolysis of dried Açaí (*Euterpe oleracea,* Mart) seeds at 450°C and 1.0 atm are shown in **Figures 5** and **6**, respectively. SEM was applied to investigate changes on the vegetal surface structure during the pyrolysis process. By comparison of SEM images of açaí seeds *in nature* and after pyrolysis, it can be observed for açaí seeds *in nature* that an aggregate, amorphous

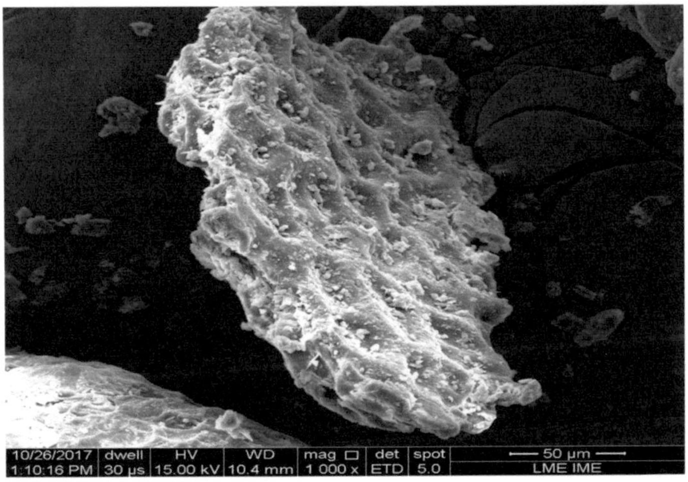

Figure 5. SEM of Açaí (*Euterpe oleracea,* Mart) seeds *in nature.*

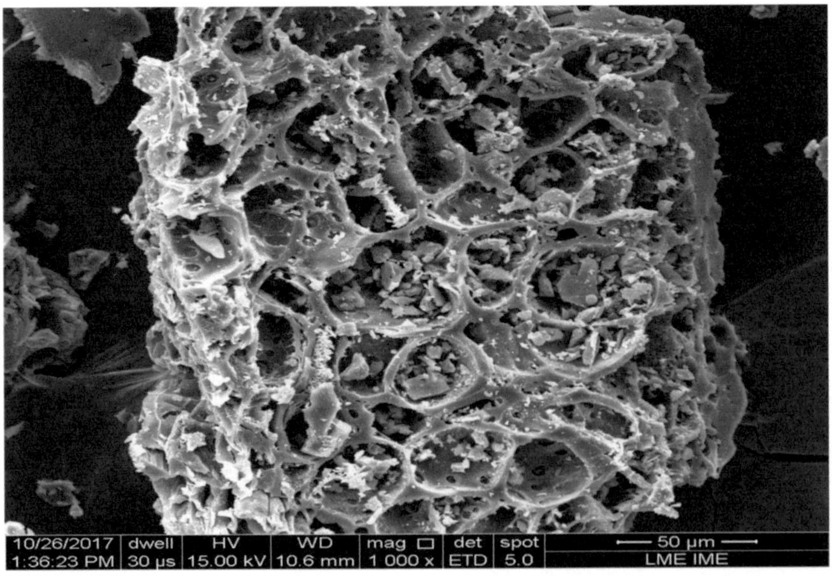

Figure 6. SEM of Açaí (*Euterpe oleracea,* Mart) seeds coke after pyrolysis at 450°C and 1.0 atm, in pilot scale.

and homogeneous structure with irregular shapes dominates, showing the pyrolysis process had a drastic effect on the vegetal morphology, as the vegetal structure differs largely from its original microscopic characteristics, as observed in **Figure 6**, as all the plant cell walls are constituted by cavities. The pyrolysis process produced an aggregate, amorphous solid phase, heterogeneous structure with irregular shapes, being the morphology after pyrolysis completely different compared to the characteristics of original vegetal surface structure. In addition, according to **Table 7**, the carbonization grade is higher, showing that temperature has caused substantial changes on the morphological structure of açaí (*Euterpe oleracea*, Mart) seeds *in nature* by destructing and/or degrading the plant cell walls.

Table 7 illustrates the energy dispersive X-ray spectroscopy of açaí seeds *in nature* and after pyrolysis of dried Açaí (*Euterpe oleracea*, Mart) seeds at 450°C and 1.0 atm, in pilot scale. The results show that after pyrolysis the carbon content increases, while that of oxygen decreases, compared to those of açaí seeds *in nature*, showing that carbonization grade is higher. Finally, the morphological structure of açaí (*Euterpe oleracea*, Mart) seeds after pyrolysis presents open cavities caused by destruction of the plant cell walls, and may be probably used as a bio-adsorbent.

3.6.2. EDX analysis of solid phase

Table 7 illustrates the energy dispersive X-ray spectroscopy of açaí seeds *in nature* and after pyrolysis of dried Açaí (*Euterpe oleracea*, Mart) seeds at 450°C and 1.0 atm by EDX technique. The results show that carbon content increases from 79.28 to 89.98% (wt.), showing a carbonization grad of 13.5%, while that of oxygen decreases from 20.71 to 6.94% (wt.) along with the pyrolysis process. The EDX also identified the presence of K and S in açaí seeds *in nature*.

3.6.3. XRD analysis of solid phase

Figure 7 shows the XRD analysis of solid phase products obtained by pyrolysis of Açaí (*Euterpe oleracea*, Mart) seeds at 450°C and 1.0 atm, in pilot scale. The results confirm the presence of

	450°C			Açaí seeds		
Chemical elements	**Mass (wt.%)**	**Atomic mass (wt.%)**	**SD**	**Mass (wt.%)**	**Atomic mass (wt.%)**	**SD**
C	89.98	93.55	4.64	79.28	83.60	4.85
O	6.94	5.41	0.54	20.71	16.39	1.50
Mg	–	–	–	–	–	–
Si	–	–	–	–	–	–
K	2.45	2.45	0.07	–	–	–
S	0.60	0.60	0.04	–	–	–
Al	–	–	–	–	–	–

SD = standard deviation.

Table 7. Percentages in *Atomic Mass* of dried Açaí (*Euterpe oleracea*, Mart) seeds *in nature* and after pyrolysis at 450°C and 1.0 atm by EDX technique.

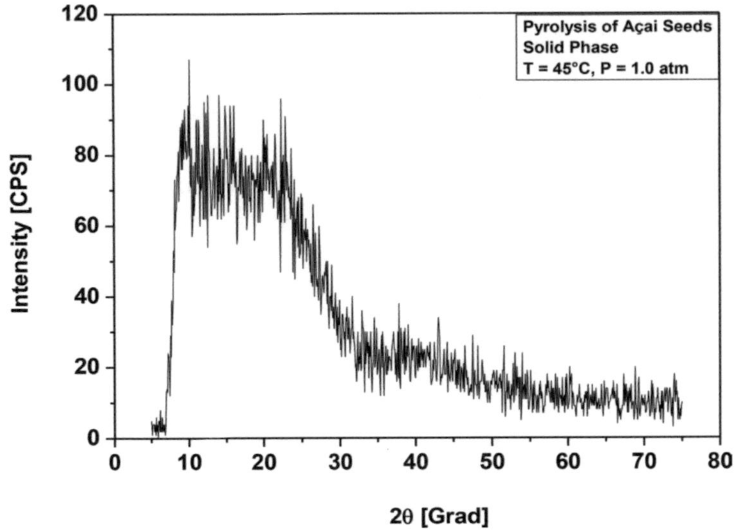

Figure 7. XRD of Açaí (*Euterpe oleracea,* Mart) seeds coke after pyrolysis at 450°C and 1.0 atm, in pilot scale.

three crystalline phases: (1) graphite (C) with a peak of high intensity (100%) on the position 2: 26.52 (ICDD: 00–025-0284); (2) cristobalite (SiO2) with peaks of medium intensity on the positions 2: 15.07 (71.53%) (ICDD: 01-077-1316); (3) quartz (SiO_2) with a peak of high intensity on the position 2: 20.40 (90.30%) (ICDD: 01-089-8940). The pyrolysis favors the formation of mineralogical phase graphite (C). This is according to the results described in Section 3.2.2, whereas a carbonization takes place during the pyrolysis process.

4. Conclusions

The experimental results show that bio-oil, gas, water phase, and coke yields were 4.38, 30.56, 29.39, and 35.67% (wt.), respectively. The bio-oil yield of 4.38% (wt.) is lower compared to similar data for bio-oil yield obtained by fast pyrolysis of forestry residues at 520°C reported in the literature [33, 34], ranging from 10 to 20% (wt.). The bio-oil density and viscosity were 1.0468 g/cm^3 and 68.34 mm^2/s, respectively, being according to similar data reported in the literature [35]. The acid value of bio-oil was 70.26 KOH/g, showing the presence of oxygenates compounds, such as carboxylic acids, phenols, cresols, ketones, and aldehydes, confirming the results reported by Oasmaa et al. [36].

The distillation of bio-oil yielded fossil fuel-like fractions (gasoline, kerosene, and light diesel) of 4.70, 28.21, and 22.35% (wt.), respectively, totalizing 55.26% (wt.), being the results according to similar studies for distillation of bio-oil reported in the literature [22, 23, 25, 37]. The acidity of distillation fractions (gasoline, kerosene, and light diesel-like) increases with increasing boiling temperature, showing a drastic decrease compared to the acidity of raw bio-oil. This i probably due to the high concentration of higher boiling-point compounds in the distillate fractions, such as p-cresol, o-cresol, guaiacol, phenol, and furans [23].

The FT-IR analysis of bio-oil identifies the presence of hydrocarbons (alkanes, alkenes, aromatic hydrocarbons, etc.) and oxygenates (phenols, cresols, carboxylic acids, ketones, furans, etc.). The bio-oil is composed by 21.52% (area) hydrocarbons (2.12% alkenes, 7.52% alkanes, 10.04% aromatic hydrocarbons, and 1.85% cycloalkenes), and 78.48% (area) oxygenates (4.06% esters, 8.52% carboxylic acids, 3.53% ketones, 35.16% phenols, 20.52% cresols, 5.75% furans, and 0.91% aldehydes). The presence of carboxylic acids, as well as phenols and cresols, but not only, confers the high acidity of bio-oil, as described in **Table 3**.

The pyrolysis process produced an aggregate, amorphous solid phase, heterogeneous structure with irregular shapes, being the morphology after pyrolysis completely different compared to the characteristics of original vegetal surface structure. In addition, the temperature has caused substantial changes on the morphological structure of açaí (*Euterpe oleracea*, Mart) seeds *in nature* by destructing and/or degrading the plant cell walls. The results of EDX show that carbon content increases from 79.28 to 89.98% (wt.), showing a carbonization grad of 13.5%, while that of oxygen decreases from 20.71 to 6.94% (wt.) along with the pyrolysis process. The results of EDX confirm the presence of three crystalline phases: (1) graphite (C); (2) cristobalite (SiO_2); (3) quartz (SiO_2), being graphite the peak of high intensity (100%) on the position 2: 26.52 (ICDD: 00-025-0284). The pyrolysis favors the formation of mineralogical phase graphite.

The fractional distillation makes it possible to obtain fossil fuel-like fractions (gasoline, kerosene, and light diesel) rich in hydrocarbons, based on the boiling temperature of hydrocarbons.

Author details

Douglas Alberto Rocha de Castro[1,2,3], Haroldo Jorge da Silva Ribeiro[1,2], Caio Campos Ferreira[1,2], Márcio de Andrade Cordeiro[1], Lauro Henrique Hamoy Guerreiro[1], Anderson M. Pereira[1,2,4], W. G. dos Santos[1,2,4], Marcelo Costa Santos[1,2], Fernanda B. de Carvalho[3], Jose Otavio Carrera Silva Junior[3], R. Lopes e Oliveira[5], Sergio Duvoisin[5], Luiz Eduardo Pizarro Borges[6] and Nélio Teixeira Machado[1,2]*

*Address all correspondence to: machado@ufpa.br

1 Laboratory of Separation Processes and Applied Thermodynamic (TERM@), Faculty of Chemical Engineering-UFPA, Belém, Pará, Brazil

2 Graduate Program of Natural Resources Engineering-UFPA, Belém, Pará, Brazil

3 Laboratory of Pharmaceutical and Cosmetic Research and Development, Faculty of Pharmacy-UFPA, Belém, Pará, Brazil

4 Faculty of Agricultural Sciences-UFAM, Coroado, Brazil

5 Faculty of Chemical Engineering-UEA, Manaus, Amazonas, Brazil

6 Laboratory of Catalyst Preparation and Catalytic Cracking, Section of Chemical Engineering-IME, Rio de Janeiro, RJ, Brazil

References

[1] Jones DL. Palms: Throughout the world. Washington: Smithsonian Institution; 1995. 410p

[2] CONAB – Companhia Nacional de Abastecimento. Açaí, Crop 2015/2016. 3rd Survey; September 2016

[3] IBGE – Instituto Brasileiro de Geografia e Estatística. Produção da Extração Vegetal e da Silvicultura, Rio de Janeiro–RJ. Vol. 292014. pp. 1-56. http://biblioteca.ibge.gov.br/visualizacao/periodicos/74/pevs_2014_v29.pdf

[4] SEDEME – Secretaria de Estado de Desenvolvimento Econômico. Mineração e Energia. Available from: http://para2030.com.br/caroco-de-acai-pode-virar-fonte-de-energia

[5] Rodriguez-Zúniga UF, Lemo V, Farinas CS, Bertitucci N, Couri S. Evaluation of agroindustrial residues as substrates for cellulolytic enzymes production under solid state fermentation. In: 7SBPMat, 2008, Guarujá; Brazilian MRS Meeting, 2008, Guarujá. Abstracts. Rio de Janeiro: SBPMat; 2008

[6] Cordeiro MA. Estudo da hidrólise enzimática do caroço de açaí (*Euterpe oleracea* Mart) para a produção de etanol [Dissertação de Mestrado]. Brazil: Programa de Pós-Graduação em Engenharia Química, UFPA. Marcio de Andrade Cordeiro; 2016

[7] da Fonseca TRB, de Amorim Silva T, Alecrim MM, da Cruz Filho RF, Teixeira MFS. Cultivation and nutritional studies of an edible mushroom from North Brazil. African Journal of Microbiology Research. 2015;**9**(30):1814-1822

[8] Kababacknik A, Roger H. Determinação do poder calorífico do caroço do açaí em três distintas umidades. In: 38th Congresso Brasileiro de Química; São Luiz-MA-Brazil; 1998

[9] Altman RFA. O Caroço de açaí (*Euterpe oleracea*, Mart). Boletim Técnico do Instituto Agronômico do Norte. Belém-Pa, Brasil. 1956;**31**:109-111

[10] Stöcker M. Biofuels and biomass-to-liquid fuels in the biorefinery: Catalytic conversion of lignocellulosic biomass using porous materials. Angewandte Chemie International Edition. 2008;**47**(48):9200-9211

[11] Santos ALF, Martins DU, Iha OK, Ribeiro RAM, Quirino RL, Suarez PAZ. Agro-industrial residues as low-price feedstock for diesel-like fuel production by thermal cracking. Bioresource Technology. 2010;**101**:6157-6162

[12] Özçimen D, Ersoy-Meriçboyu A. Characterization of biochar and bio-oil samples obtained from carbonization of various biomass materials. Renewable Energy. 2010;**35**(6): 1319-1324

[13] Nelson DL, Cox MM. Lehninger Principles of Biochemistry. 5th ed. New York, NY: Freeman; 2008. ISBN: 978-0-7167-7108-1

[14] Roberts KG, Gloy BA, Joseph S, Scott NR, Lehmann J. Life cycle assessment of biochar systems: Estimating the energetic, economic, and climate change potential. Environmental Science & Technology. 2010;**44**(2):827-833

[15] Guerreiro LHH, Castro DAR, Rodrigues LF, Santos MC, Ferreira CC, Machado NT. Investigação do processo de destilação de bioóleo de sementes de Açaí (*Euterpe oleracea*) obtido via pirólise em escala piloto. In: XV Encontro de Profissionais da Química da Amazônia-XV EPQA, 2017, Belém. Vol. 1; XV Encontro de Profissionais da Química da Amazônia-XV EPQA; 2017. p. 1

[16] Guerreiro LHH, Castro DAR, Ribeiro HJS, Ferreira CC, Santos MC, Machado NT. Investigação do processo de destilação de bioóleo de sementes de Açaí (*Euterpe oleracea*) obtido via pirólise em escala piloto. In: 9 Congresso Brasileiro de Pesquisa e Desenvolvimento em Petróleo e Gás, Maceió; Vol. 1; 2017. p. 1

[17] Guo Z, Wang S, Zhu Y, Luo Z, Cen K. Separation of acid compounds for refining biomass pyrolysis oil. Journal of Fuel Chemistry and Technology. 2009;**7**(1):49-52

[18] Guo X, Wang S, Guo Z, Liu Q, Luo Z, Cen K. Pyrolysis characteristics of bio-oil fractions separated by molecular distillatio. Applied Energy. 2010;**87**(9):2892-2898

[19] Guo Z, Wang S, Gu Y, Xu G, Li X, Luo Z. Separation characteristics of biomass pyrolysis oil in molecular distillation. Separation and Purification. 2010;**76**(1):52-57

[20] Christensen ED, Chupka GM, Luecke J, Smurthwaite T, Alleman TL, Iisa K, Franz JA, Elliott DC, McCormick RL. Analysis of oxygenated compounds in hydrotreated biomass fast pyrolysis oil distillate fractions. Energy & Fuels. 2011;**25**(11):5462-5471

[21] Li S-Q, Yao Q, Chi Y, Yan J-H, Cen K-F. Pilot-scale pyrolysis of scrap tires in a continuous rotary kiln reactor. Industrial and Engineering Chemistry Research. 2004;**43**:5133-5145

[22] Zheng J-L, Wie Q. Improving the quality of fast pyrolysis bio-oil by reduced pressure distillation. Biomass and Bioenergy. 2011;**35**:1804-1810

[23] Zhang X-S, Yang G-X, Jiang H, Liu W-J, Ding H-S. Mass production of chemicals from biomass-derived oil by directly atmospheric distillation coupled with co-pyrolysis. Scientific Reports. 2013;**3**(1120):1-7

[24] Sarma AK, Konwer D. Feasibility studies for conventional refinery studies for conventional refinery distillation with a (1:1) w/w of a biocrude blend with petroleum crude obiocrude blend with petroleum crude oil. Energy & Fuels. 2005;**19**(4):1755-1758

[25] Elkasabi Y, Mullen CA, Boateng AA. Distillation and isolation of commodity chemicals from bio-oil made by tail-gas reactive pyrolysis. ACS Sustainable Chemistry & Engineering. 2014;**2**:2042-2052

[26] Kanaujia PK, Naik DV, Tripathi D, Singh R, Poddar MK, Siva Kumar Konathala LN, Sharma YK. Pyrolysis of *Jatropha curcas* seed cake followed by optimization of liquid–liquid extraction procedure for the obtained bio-oil. Journal of Analytical and Applied Pyrolysis. 2016;**118**:202-224

[27] Vispute TP, Huber GW. Production of hydrogen, alkanes and polyols by aqueous phase processing of wood-derived pyrolysis oils. Green Chemistry. 2009;**11**:1433-1445

[28] Song Q, Nie J, Ren M, Guo Q. Effective phase separation of biomass pyrolysis oils by adding aqueous salt solutions. Energy & Fuels. 2009;**23**:3307-3312

[29] Buffiere P, Loisel D. Dosage des Fibres Van Soest. Narbonne: Weened, Laboratoire de Biotechnologie de l'Environnement, INRA; 2007. pp. 1-14

[30] Tappi Standards. Acid-Insoluble Lignin in Wood and Pulp. Tappi Method T 222 om-06. Atlanta, GA: Tappi Press; 2006

[31] da Mota SAP, Mancio AA, Lhamas DEL, de Abreu DH, da Silva MS, dos Santos WG, et al. Production of green diesel by thermal catalytic cracking of crude palm oil (*Elaeis guineensis* Jacq) in a pilot plant. Journal of Analytical and Applied Pyrolysis. 2014;**110**:1-11

[32] Machado NT, de Castro DAR, Queiroz LS, Santos MC, da Costa CEF. Production and characterization of energy materials with adsorbent properties by hydrothermal processing of corn stover with subcritical H_2O. Journal of Applied Solution Chemistry and Modeling. 2016;**5**:117-130

[33] Mohan D, Pittman CU Jr, Steelee PH. Pyrolysis of wood/biomass for bio-oil: A critical review. Energy & Fuels. 2006;**20**:848-889

[34] Oasmaa A, Kuoppala E, Gust S, Solantausta Y. Fast pyrolysis of forestry residue. 1. Effect of extractives on phase separation of pyrolysis liquids. Energy & Fuels. 2003;**17**(1):1-12

[35] Czernik S, Bridgwater AV. Overview of applications of biomass fast pyrolysis oil. Energy & Fuels. 2004;**18**:590-598

[36] Oasmaa A, Elliott DC, Korhonen J. Acidity of biomass fast pyrolysis bio-oils. Energy & Fuels. 2010;**24**(12):6548-6554

[37] Boateng AA, Mullen CA, Goldberg N, Hicks KB. Production of bio-oil from Alfalfa stems by fluidized-bed fast pyrolysis. Industrial and Engineering Chemistry Research. 2008;**47**:4115-4122

[38] Ferreira CC, Costa EC, de Castro DAR, Pereira MS, Mâncio AA, Santos MC, et al. Deacidification of organic liquid products by fractional distillation in laboratory and pilot scales. Journal of Analytical and Applied Pyrolysis. 2017;**127**:468-489

[39] Seshadri KS, Cronauer DC. Characterization of coal-derived liquids by 13^C N.M.R. and FT-IR Spectroscopy. Fuel. 1983;**62**:1436-1444